Study Guide to Accompany

UNDERSTANDING STATISTICS
IN THE BEHAVIORAL SCIENCES

FOURTH EDITION

ROBERT R. PAGANO, PH.D
UNIVERSITY OF PITTSBURGH

PREPARED BY
WILLIAM C. FOLLETTE, PH.D
UNIVERSITY OF NEVADA

ROBERT R. PAGANO, PH.D
UNIVERSITY OF PITTSBURGH

WEST PUBLISHING COMPANY
MINNEAPOLIS/ST. PAUL   NEW YORK   LOS ANGELES   SAN FRANCISCO

WEST'S COMMITMENT TO THE ENVIRONMENT

In 1906, West Publishing Company began recycling materials left over from the production of books. This began a tradition of efficient and responsible use of resources. Today, up to 95% of our legal books and 70% of our college texts and school texts are printed on recycled, acid-free stock. West also recycles nearly 22 million pounds of scrap paper annually—the equivalent of 181,717 trees. Since the 1960s, West has devised ways to capture and recycle waste inks, solvents, oils, and vapors created in the printing process. We also recycle plastics of all kinds, wood, glass, corrugated cardboard, and batteries, and have eliminated the use of Styrofoam book packaging. We at West are proud of the longevity and the scope of our commitment to the environment.

Production, Prepress, Printing and Binding by West Publishing Company.

 TEXT IS PRINTED ON 10% POST CONSUMER RECYCLED PAPER    PRINTED WITH SOY INK™

Contents

PREFACE

Statistics uses probability, logic and mathematics as a way of determining whether or not observations made in the real world or laboratory are due to random happenstance or perhaps due to an orderly effect one variable has on another. Separating happenstance, or chance, from cause and effect is the task of science, and statistics is a tool to accomplish that end. Occasionally, data will be so clear that the use of statistical analysis isn't necessary. Occasionally, data will be so garbled that no statistics can meaningfully be applied to it to answer any reasonable question. But we will demonstrate that most often statistics is useful in identifying whether it is legitimate to conclude that an orderly effect has occurred.

It is useful to try to think of statistics as a means of learning a new set of problem solving skills. You will learn new ways to ask questions, new ways to answer them, and a more sophisticated way of interpreting the data you read about in texts, journals, and the newspapers.

Each chapter in this study guide begins by outlining the corresponding chapter in the textbook. Next comes a section called Concept Review which is a "fill in the blanks" exercise. This is a useful way to make sure you understand the basic ideas in the chapter. Following the concept review, there are sets of exercises which are designed to help you practice the mechanics of solving the problems, recognize what tests to apply to a real world problem and notice any changes in computational procedures required for special cases. A set of True-False questions follow the exercise section. Finally, there is a Self-Quiz which is presented in a multiple choice format.

In our experience there is a direct, positive relationship between working problems and doing well on this material. We encourage you to work as many problems as you possibly can. When using calculators and computers, there can be a tendency to press the keys and read the answer without having the slightest idea where the answer came from. We hope you won't fall into this trap.

One final point - comparing your answers to ours. For most of the problems we have used a hand calculator or computer to find the solutions. Depending on how many decimal places you carry your intermediate calculations, you may get slightly different answers than we do. In most cases we have used full calculator accuracy for intermediate calculations (at least five decimal places). In general, you should carry all intermediate calculations to at least two more decimal places than the number of decimal places in the rounded final answer. For example, if you intend to round the final answer to two decimal places, than you should carry all intermediate calculations to at least 4 decimal places. If you follow this policy and your answer does not agree with ours, then you have probably made a calculation error.

Many students have commented that they have found the first three editions of the study guide quite useful. This fourth edition is very much like the first three, except that it contains additional material to cover the new material in the textbook. We hope you will find this fourth edition helpful in understanding this difficult subject matter.

William C. Follette
Robert R. Pagano

Pittsburgh, Pennsylvania
January, 1994

1 | STATISTICS AND SCIENTIFIC METHOD

CHAPTER OUTLINE

I. Methods of Acquiring Knowledge

A. <u>Authority</u>. One accepts information as being true because someone who is supposed to know tells you something is true.

B. <u>Rationalism</u>. This method uses reason alone to arrive at knowledge. One analyzes a situation and draws logical conclusions based on the information at hand. The conclusion is not tested empirically to determine if it is correct.

C. <u>Intuition</u>. This is a sudden insight that springs into consciousness all at once as a whole.

D. <u>Scientific method</u>. This method uses reasoning and intuition as a means of formulating an idea of what is true but then relies on objective assessment to verify or deny the validity of the idea.

　　1. Idea formed and hypothesis made.

　　2. Experiment designed.

　　3. Data collected and analyzed using statistics.

　　4. Hypothesis confirmed, denied or modified.

II. Scientific Research

A. Observational studies. In this research there is no direct experimental manipulation of variables. This technique employs naturalistic observation of events in their real world environment.

   1. Correlation. A type of observation where the relationship between two variables is inferred.

   2. Parameter estimation. This is when an investigator tries to determine the actual characteristics of the population, based on measuring a subset of the population.

B. True experiments. The investigator attempts to determine if changes in one variable produce changes in another. In both observational studies and true experiments, statistical analysis is usually employed.

C. Statistical Analysis.

   1. Descriptive statistics. Analysis is conducted to describe the obtained data.

   2. Inferential statistics. Analysis is conducted to make inferences about a population using data obtained from the sample.

CONCEPT REVIEW

There are at least four methods of acquiring knowledge.

When one accepts something as true because of tradition

or because a person of distinction says it is true, one is using

the method of (1) _____. When one uses the rules of   (1) authority

reason and logic alone to arrive at truth, this is called

(2) _____. Sometimes a sudden insight provides a means   (2) rationalism

of acquiring knowledge. This is referred to as (3) _____.   (3) intuition

The most refined method of acquiring knowledge is by us-

ing the (4) _____ (5) _____. The scientific method   (4) scientific
                                                                   (5) method
uses reasoning and intuition but it relies on (6) _____   (6) objective

(7) _____ using (8) _____ to confirm or refute

hypotheses. Scientific research utilizes two basic research

methods to acquire knowledge. These are (9) _____

studies and (10) _____ experiments. In observational

studies the experimenter (11) <u>does/does not</u> actively manip-

ulate variables. This method employs (12) _____ obser-

vation, (13) _____ estimation, and (14) _____ studies.

In many of these studies, the goal is to accurately (15) _____

a situation or relationship.

In true experiments the goal is to determine whether changes

in one variable (16) _____ (17) _____ in another variable.

The variable which the experimenter has control over and ma-

nipulates is the (18) _____ variable. The variable that is ob-

served for changes is called the (19) _____ variable.

It is usually not possible to collect data on the entire

(20) _____ so a subset, called a (21) _____ is studied. It

is crucial that this sample is a (22) _____ sample. This helps

assure that the laws of (23) _____ apply to the data and that

the sample is (24) _____ of the population.

The study of statistics is divided into two areas. The first is

(25) _____ (26) _____ which describes and characterizes

data. The second area, which allows us to use sample data to

infer conclusions about the population is called (27) _____

(28) _____.

(7) assessment
(8) experiment-
     ation

(9) observational

(10) true

(11) does not

(12) naturalistic

(13) parameter
(14) correlational
(15) describe

(16) produce
(17) changes

(18) independent

(19) dependent

(20) population
(21) sample
(22) random

(23) probability

(24) representative

(25) descriptive
(26) statistics

(27) inferential

(28) statistics

The complete set of individuals, objects, or scores that the investigator is interested in is called the (29) _____. The sub-set of the population which the experimenter studies is called a (30) _____. Properties or characteristics of some event, object or person which can take on different values are called (31) _____. Measurements made on the dependent variable are called (32) _____. Data which have not yet been anal-yzed or summarized are referred to as (33) _____ data or (34) _____ scores. A (35) _____ is a measure of a a sample characteristic. A (36) _____ is a measure of a population characteristic. A statistic and a parameter are re-lated concepts. A statistic refers to (37) _____ data, while a parameter refers to the entire (38) _____.

(29) population

(30) sample

(31) variables

(32) data

(33) raw

(34) original
(35) statistic
(36) parameter

(37) sample

(38) population

EXERCISES

1. What is the difference between rationalism and the scientific method?

2. A social scientist measured the number of years of education of a random group of people and asked them their income. The scientist then found that the longer the education, the higher the income.

   a. What kind of study was this?
   b. Do you think that the scientist can rightfully infer that higher education causes higher income?

3. A scientist takes a random sample of 50 citizens in a city and asks them their age. The scientist then calculates the average age and uses that as an estimate of the average age of all the citizens in the city. What is this process called?

4. Another scientist is interested in parameter estimation. When asked to estimate the average income in a city he went to the Millionaire's Club and asked each person their average income. What was wrong with that process?

5. Consider yourself and describe 3 populations to which you belong?

6. How often are population parameters known?

7. Give three examples of a variable applying to people.

8. In an experiment, a scientist is interested in the effect of a drug on reaction time. He gives various doses of the drug and then measures blood levels of the drug and reaction time in a driving task. What is the independent variable? The dependent variable?

9. If someone asked you the average height of everyone in your family and you actually measured each of the people in your family to calculate the average, have you calculated a statistic or parameter?

10. A student hears a lecture on the harmful effects of drug X on intellectual functioning. The student then tells his friend that drug X is harmful. What method of inquiry has the student relied upon to learn about drug X?

Answers: 1. Both use reason to arrive at knowledge, but in addition, scientific method uses objective assessment to test the validity of an idea. **2a.** correlational study **2b.** No, just because two variables are related, doesn't mean one causes the other. (We'll talk more about this topic in Chapter 7. **3.** Inferential statistics. In this case it is parameter estimation in which the parameter being estimated is the average age of the citizens in the city. **4.** The sample was not random, so the results can only be generalized to the Millionaire's Club. **5.** Any three will do. Some that may apply are college students, males/females, residents of your city, citizens of a country, members of a particular religious group, etc. **6.** Rarely, usually a sample is used to estimate the parameter. **7.** Age, hair color, number of teeth missing, visual acuity, income, religion, race, sex, social security number, etc. **8.** The independent variable is the drug dosage and the dependent variable is reaction time. **9.** population parameter. **10.** authority.

TRUE-FALSE QUESTIONS

T F   1. Intuition is not a valid way of developing research hypotheses.

T F   2. The main difference between rationalism and the scientific method is that the scientific method relies on objective assessment to test the idea.

T F   3. In observational studies the researcher manipulates the experiment and observes the results closely.

T F   4. One task of research is to study a subset of the population and try to make inferences about the whole population.

T F   5. Naturalistic observation is rarely used by scientists because it does not make use of the scientific method.

T F   6. Descriptive statistics are used to make inferences about the population.

T F   7. In a study of caloric intake on weight loss, the number of pounds lost (or gained) is the independent variable.

T F   8. In inferential statistics random sampling is regarded as desirable but not essential.

T F   9. If one measured the height of all the women at a university and took an average of the data, the university women would be a sample of the population of all university women in the world.

T F 10. When data has not yet been analyzed, this is called raw data.

Answers: 1. F   **2.** T   **3.** F   **4.** T   **5.** F   **6.** F   **7.** F   **8.** F   **9.** T   **10.** T.

SELF-QUIZ

1. When one analyzes data based on a sample, one calculates a _____.

    a. parameter
    b. variable
    c. constant
    d. statistic

2. Mathematical methods used to draw tentative conclusions about a population based on sample data are referred to as _____.

    a. descriptive statistics
    b. sample statistics
    c. inferential statistics
    d. random sampling
    e. magic

3. The variable which the experimenter manipulates is called the _____.

    a. independent variable
    b. dependent variable
    c. constant variable
    d. experimental variable

4. In acquiring knowledge the method which employs logic, reasoning and objective assessment is referred to as _____.

   a. the method of authority
   b. intuition
   c. rationalism
   d. scientific method

5. From which of the following studies can one most reasonably determine cause and effect?

   a. correlational study
   b. true experiment
   c. naturalistic observation
   d. all the above equally well

6. In inferential statistics the object is usually to generalize from a _____ to a _____.

   a. data; variable
   b. sample; population
   c. population; sample
   d. constant; variable

7. For a list of data the lowest and highest score values are examples of _____ statistics.

   a. inferential
   b. sample
   c. population
   d. descriptive

8. To avoid unknown, systematic factors which may bias the results of an experiment, the experimenter should select a _____ sample from a population and use controlled conditions.

   a. wise
   b. small
   c. random
   d. single

9. Descriptive statistics are helpful in _____ and _____ raw data.

   a. generalizing; inferring
   b. organizing; tenderizing
   c. confusing; confounding
   d. summarizing; characterizing

10. I am certain I will find this course _____.

    a. fun
    b. fatal
    c. interesting
    d. insufficient data to answer question

Answers: 1. d **2.** c **3.** a **4.** d **5.** b **6.** b **7.** d **8.** c **9.** d **10.** Give yourself credit for any answer other than b.

2 BASIC MATHEMATICAL AND MEASUREMENT CONCEPTS

CHAPTER OUTLINE

I. **Study Hints for the Student**

   A Review basic algebra but don't be afraid that the mathematics will be too hard.

   B. Become very familiar with the notations in the book.

   C. Don't fall behind. The material in the book is cumulative and getting behind is a bad idea.

   D. Work problems!

II **Mathematical Notation**

   A. The symbols X (capital letter X) and sometimes Y will be used as symbols to represent variables measured in the study.

   1. For example, X could stand for age, or height, or IQ in any given study.

   2. To indicate a specific observation a subscript on X will be used; e.g., X_2 would mean the second observation of the X variable.

B. The summation sign (Σ) is used to indicate the fact that the scores following the summation sign are to be added up. The notations above and below the Σ sign are used to indicate the first and last scores to be summed.

C. Summation Rules.

1. The sum of the values of a variable plus a constant is equal to the sum of the values of the variable plus N times the constant.

   In equation form:

   $$\sum_{i=1}^{N} (X_i + a) = \sum_{i=1}^{N} X_i + Na$$

2. The sum of the values of a variable minus a constant is equal to the sum of the variable minus N times the constant.

   In equation form:

   $$\sum_{i=1}^{N} (X_i - a) = \sum_{i=1}^{N} X_i - Na$$

3. The sum of a constant times the values of a variable is equal to the constant times the sum of the values of the variable.

   In equation form:

   $$\sum_{i=1}^{N} aX_i = a\sum_{i=1}^{N} X_i$$

4. The sum of a constant divided into the values of a variable is equal to the constant divided into the sum of the values of the variable.

   In equation form:

   $$\sum_{i=1}^{N} (X_i/a) = \left(\sum_{i=1}^{N} X_i\right)/a$$

III. Measurement Scales.

A. All measurement scales have one or more of the following three attributes.

1. Magnitude.

2. Equal intervals between adjacent units.

3. Absolute zero point.

B. The <u>nominal scale</u> is the lowest level of measurement. It is more qualitative than quantitative. Nominal scales are comprised of elements which have been classified as belonging to a certain category. For example, whether someone's sex is male or female. Can only determine whether A = B or A ≠ B.

C. <u>Ordinal scales</u> possess a relatively low level of the property of magnitude. The rank order of people according to height is an example of an ordinal scale. One does not know how much taller the first rank person is over the second rank person. Can determine whether A > B, A = B or A < B.

D. <u>Interval scales.</u> This scale possesses equal intervals, magnitude, but no absolute zero point. An example is temperature measured in degrees Celsius. What is called zero is actually the freezing point of water, not absolute zero. Can do same determinations as ordinal scale, plus can determine if A - B = C - D, A - B > C - D or A - B < C - D.

E. <u>Ratio scales</u>. These scales have the most useful characteristics since they possess attributes of magnitude, equal intervals, and an absolute zero point. All mathematical operations can be performed on ratio scales. Examples include height measured in centimeters, reaction time measured in milliseconds.

IV. Additional Points Concerning Variables

A. <u>Continuous variables</u>. This type can be identified by the fact that they can theoretically take on an infinite number of values between adjacent units on the scale. Examples include length, time and weight. For example, there are an infinite number of possible values between 1.0 and 1.1 centimeters.

B. <u>Discrete variables</u>. In this case there are no possible values between adjacent units on the measuring scale. For example, the number of people in a room has to be measured in discrete units. One cannot reasonably have 6 1/2 people in a room.

C. <u>Limits</u>. All measurements on a continuous variable are approximate. They are limited by the accuracy of the measurement instrument. When a measurement is taken one is actually specifying a range of values and calling it a specific value. The real limits of a continuous variable are those values which are above and below the recorded value by 1/2 of the smallest measuring unit of the scale (e.g., the real limits of 100°C are 99.5° C and 100.5° C, when using a thermometer with accuracy to the nearest degree).

D. <u>Significant figures</u>. The number of decimal places in statistics is established by tradition. The advent of calculators has made carrying out laborious calculations much less cumbersome. Because solutions to problems often involve a large number of intermediate steps, small rounding inaccuracies can become large errors. Therefore, the more decimals carried in intermediate calculations, the more accurate is the final answer. It is standard practice to carry to one or more decimal places in intermediate calculations than you report in the final answer.

E. <u>Rounding</u>. If the remainder beyond the last digit is greater than 1/2 add one to the last digit. If the remainder is less than 1/2 leave the last digit the same. If the remainder is equal to 1/2 add one to the last digit if it is an odd number, but if it is even, leave it as it is.

CONCEPT   REVIEW

In statistics we often let symbols stand for variables. Generally

we will let the letter (1) _____ stand for a variable such as height,   (1) X

IQ or reaction time. We will also sometimes use the letter Y. We

will use (2) _____ on X to indicate a specific case. For example,   (2) subscripts

one would designate the third value of IQ in a set of scores as

(3) _____. One would indicate the nth score as (4) _____.   (3) X_3
(4) X_n

One very common operation is to sum all the scores. To indicate

this operation, we use the symbol (5) _____, which is the   (5) Σ

Greek capital letter (6) _____. Summing a set of scores   (6) sigma

simply means to (7) _____. Algebraically, we can write a   (7) add them

mathematical sentence which can easily be translated into English.

The symbols $\sum_{i=1}^{N} X_i$ can be translated as meaning (8) _____   (8) sum

of the (9) _____ variable from i = (10) _____ to (11) _____.   (9) X
(10) 1
(11) N

The notations above and below the summation sign designate

which (12) _____ to include in the summation. The term   (12) scores

(13) _____ the summation sign tells us the first score in the   (13) below

summation. The term above the Σ designates the (14) _____   (14) last

score. If the summation is over all the scores (from 1 to N) we can

abbreviate the mathematical sentence to (15) _____. If one   (15) ΣX

wanted to indicate summing the 3rd, 4th, and 5th scores, one

could show this mathematically as (16) _____.   (16) $\sum\limits_{i=3}^{5} X_i$

The expression for summing all the squared X scores is

(17) _____. The notation for summing all the X scores   (17) ΣX^2

and squaring the final sum is (18) _____.   (18) $(\Sigma X)^2$

There are four types of measurement scales. Measurements

which classify objects into groups with names are called

(19) _____ scales. This is primarily a measurement which names   (19) nominal

objects. Nominal scales are generally (20) _____ not   (20) qualitative

quantitative.

The (21) _____ scale is the next higher level of   (21) ordinal

measurement. Ordinal scales possess a low level of the

property of (22) _____. A common application of ordinal   (22) magnitude

scales is to (23) _____ order objects. The rank order tells   (23) rank

which object has (24) _____ or (25) _____ of a given   (24) more
(25) less

attribute but not how much more or less. Thus, it can be said that

ordinal scales do not have (26) _____ intervals between   (26) equal

adjacent units.

A scale which possess both properties of magnitude and equal

intervals between adjacent units is the (27) _____ scale.   (27) interval

The difference between an interval scale and a (28) _____   (28) ratio scale is that in addition, a ratio scale possesses an (29) _____   (29) absolute

zero point. The Celsius temperature scale would be an example

of an (30) _____ scale because the zero point on that scale   (30) interval

is not at absolute zero. With interval scales, we can tell whether

two objects possess more, less or the same amount of a given

attribute as well as (31) _____ of the attribute. Addition,   (31) how much
                                                                               more or less
subtraction, multiplication, division and doing ratios are per-

mitted with (32) _____ scales   (32) ratio

Earlier we defined a variable as a property or characteristic of

something which can take more than one (33) _____. Variables   (33) value

can be either continuous or discrete. A continuous variable can

have an (34) _____ number of values between adjacent units.   (34) infinite

Discrete variables have (35) _____ possible values between   (35) no

adjacent units. Length measured in centimeters is an example

of a (36) _____ variable. The number of siblings in a specific   (36) continuous

family is an example of a (37) _____ variable. Because of their   (37) discrete

nature, measures on continuous variables are (38) _____.   (38) approximate

We aren't able to measure the exact value for a continuous

variable but we can specify the (39) _____ limits for a value.   39) real

The real limits are those values which are above and below

the recorded value by one-half of the (40) _____ measuring   (40) smallest

unit of the scale. If we were measuring distance to the nearest mile

and observed that a trip was 16 miles long, the lower real limit

is (41) _____ miles and the upper real limit is (42) _____   (41) 15.5
                                                                       (42) 16.5

miles.

    In rounding answers in statistics we use the rule: If the remainder beyond the last digit is greater than 1/2, (43) _____ one to the last digit. If the remainder is less than 1/2, (44) _____ the last digit as it is. If the remainder is equal to 1/2, add one to the last digit if it is an (45) _____ number, but if it is (46) _____, leave it as it is. Thus, rounding 3.14159 to four decimals gives (47) _____. Rounding 3.5 to a whole number gives (48) _____. Rounding 3.650 to one decimal gives (49) _____. Rounding 2.5001 to the nearest whole number gives (50) _____.

(43) add

(44) leave

(45) odd

(46) even

(47) 3.1416

(48) 4

(49) 3.6

(50) 3

EXERCISES

1. Consider the following sample scores for the variable weight:

$X_1 = 145$, $X_2 = 160$, $X_3 = 110$, $X_4 = 130$, $X_5 = 137$, $X_6 = 172$, **AND** $X_7 = 150$

    a. What is the value for $\sum X$?

    b. What is the value for $\sum_{i=3}^{6} X_i$?

    c. What is the value for $\sum X^2$?

    d. What is the value for $(\sum X)^2$?

    e. What is the value for $\sum (X + 4)$?

    f. What is the value for $\sum X - 140$?

    g. What is the value for $\sum (X - 140)$?

2.   Round the following to one decimal place.

    a.   25.15
    b.   25.25
    c.   25.25001
    d.   25.14999
    e.   25.26

3.   State the real limits for the following values of a continuous variable.

    a.   100 (smallest unit of measurement is 1)
    b.   1.35 (smallest unit of measurement is 0.01)
    c.   29.1 (smallest unit of measurement is 0.1)

4.   Indicate whether the following variables are discrete or continuous.

    a.   The age of an experimental subject.
    b.   The number of ducks on a pond.
    c.   The reaction time of a subject on a driving task.
    d.   A rating of leadership on a 3 point scale.

5.   Identify which type of measurement scale is involved for the following:

    a.   The sex of a child.
    b.   The religion of an individual
    c.   The rank of a student in an academic class.
    d.   The attitude score of a subject on a prejudice inventory.
    e.   The time required to complete a task.
    f.   The rating of a task as either "easy," "mildly difficult," or "difficult."

Answers: **1a.** 1004 **1b.** 549 **1c.** 146,478 **1d.** 1,008,016 **1e.** 1032 **1f.** 864 **1g.** 24 **2a.** 25.2 **2b.** 25.2 **2c.** 25.3 **2d.** 25.1 **2e.** 25.3 **3a.** 99.5 - 100.5 **3b.** 1.345 - 1.355 **3c.** 29.05 - 29.15 **4a.** continuous **4b.** discrete **4c.** continuous **4d.** discrete **5a.** nominal **5b.** nominal **5c.** ordinal **5d.** interval (assuming equal interval) or ordinal (if equal interval assumption is unreasonable) **5e.** ratio **5f.** ordinal.

TRUE-FALSE QUESTIONS

T F   1.   All scales possess magnitude, equal intervals between adjacent units, and an absolute zero point.

T F   2.   Nominal scales can be used either qualitatively or quantitatively.

T F   3. With an ordinal scale one cannot be certain that the magnitude of the distance between any two adjacent points is the same.

T F   4. With the exception of division, one can perform all mathematical operations on a ratio scale.

T F   5 The average number of children in a classroom is an example of a discrete variable.

T F   6 When a weight is measured to 1/1000th of a gram, that measure is absolutely accurate.

T F   7. If the quantity $\sum X = 400.3$ for N observations, then the quantity $\sum X$ will equal 40.03 if each of the original observations is multiplied by 0.1.

T F   8. One generally has to specify the real limits for discrete variables since they cannot be measured accurately.

T F   9. The symbol $\sum$ means square the following numbers and sum them.

T F 10. Rounding 55.55 to the nearest whole number gives 55.

Answers: 1. F   **2.** F   **3.** T   **4.** F   **5.** F   **6.** F   **7.** T   **8.** F   **9.** F   **10.** F.

SELF-QUIZ

Refer to the following set of numbers to answer questions 1 - 6:

$$X_1 = 2, X_2 = 4, X_3 = 6, X_4 = 10$$

1. What is the value for $\sum X$?

    a.   12
    b.   156
    c.   480
    d.   22

2. What is the value of $\sum X^2$?

    a.   156
    b.   22
    c.   480
    d.   37

3. What is the value of X_4^2?

    a.   4
    b.   6
    c.   100
    d.   10

4. What is the value of $(\Sigma X)^2$?

    a.   480
    b.   484
    c.   156
    d.   44

5. What is the value of N?

    a.   2
    b.   4
    c.   6
    d.   10

6. What is the value of $(\Sigma X)/N$?

    a. 5
    b. 4
    c. 6
    d. 5.5

7. Classifying subjects on the basis of sex is an example of using what kind of scale?

    a. nominal
    b. ordinal
    c. interval
    d. ratio
    e. bathroom

8. Number of bar presses is an example of a(n) _____ variable.

    a. discrete
    b. continuous
    c. nominal
    d. ordinal

9. Using an ordinal scale to assess leadership, which of the following statements is appropriate?

   a. A has twice as much leadership ability as B
   b. X has no leadership ability
   c. Y has the most leadership ability
   d. all of the above

10. The number of legs on a centipede is an example of a _____ scale.

   a. nominal
   b. ordinal
   c. ratio
   d. continuous

11. What are the real limits of the observation of 6 seconds (measured to the nearest second)?

   a. 5.5-6.5
   b. 5.0-7.0
   c. 5.9-6.1
   d. 6.0-6.5

12. What is 17.295 rounded to one decimal place?

   a. 17.1
   b. 17.0
   c. 17.2
   d. 17.3

13. What is the value of .05 rounded to one decimal place?

   a. .0
   b. .1
   c. .2
   d. .5

14. The symbol "Σ" means:

   a. add the scores
   b. summarize the data
   c. square the value
   d. multiply the scores

The following problems are for your own use in evaluating your skills at elementary algebra. If you do not get all the problems correct you should probably review your algebra.

15. Where $3X = 9$, what is the value of X?

    a. 3
    b. 6
    c. 9
    d. 12

16. For $X + Y = Z$, X equals _____.

    a. $Y + Z$
    b. $Z - Y$
    c. Z/Y
    d. Y/Z

17. $1/X + 2/X$ equals _____.

    a. $2/X$
    b. $3/2X$
    c. $3/X$
    d. $2/X^2$

18. What is $(4 - 2)(3 \cdot 4)/(6/3)$?

    a. 24
    b. 1.3
    c. 12
    d. 6

19. $6 + 4 \times 3 - 1$ simplified is _____.

    a. 29
    b. 48
    c. 71
    d. 17

20. $X = Y/Z$ can be expressed as _____.

    a. $Y = (Z)(X)$
    b. $X = Z/Y$
    c. $Y = X/Z$
    d. $Z = X + Y$

21. 2^4 equals _____.

    a.   4
    b. 32
    c.   8
    d. 16

22. $\sqrt{81}$ equals _____.

    a. ± 3
    b. ± 81
    c. ± 9
    d. ± 27

23. $X(Z + Y)$ equals _____.

    a. $XZ + Y$
    b. $ZX + YX$
    c. $(X)(Y)(Z)$
    d. $(Z + Y)/X$

24. $1/2 + 1/4$ equals _____.

    a. 1/6
    b. 1/8
    c. 2/8
    d. 3/4

25. X^6/X^2 equals _____.

    a. X^8
    b. X^4
    c. X^2
    d. X^3

Answers: 1. d **2.** a **3.** c **4.** b **5.** b **6.** d **7.** a **8.** a **9.** c **10.** c **11.** a **12.** d **13.** a **14.** a **15.** a **16.** b **17.** c **18.** c **19.** d **20.** a **21.** d **22.** c **23.** b **24.** d **25.** b.

3 FREQUENCY DISTRIBUTIONS

CHAPTER OUTLINE

I. **Constructing Frequency Distributions**

A. A list of rank ordered scores and their frequency of occurrence in tabular form is called a frequency distribution. When the data is rank ordered, it is easier to understand.

B. <u>Grouping data</u>. Individual scores are often grouped together into class intervals of equal width to allow one to visualize the shape of a distribution and its central tendency. This is called a frequency distribution of grouped scores.

   1. Distributions are usually divided into 10 to 20 intervals.

   2. If the intervals are too wide, too much information is lost; if too narrow, there are too many zero frequency intervals which distort the shape of the distribution.

C. Constructing a frequency distribution of grouped scores.

   1. Find the range of scores.

   2. Determine the width of each class interval (i).

   3. List the limits of each class interval, placing the interval containing the lowest score at the bottom.

4. Tally the raw scores into the appropriate class intervals.

5. Add the tallies to obtain the interval frequency.

II. Desired Frequency Distributions. It is often desirable to express data in forms other than the basic frequency distribution.

A. Relative frequency distribution. This indicates the proportion of the total number of scores which occurred in each interval.

B. Cumulative frequency distribution. This indicates the number of scores which fell below the upper real limit of each interval.

C. Cumulative percentage distribution. This indicates the percentage of scores which fell below the upper real limit of each interval. To convert cumulative frequencies to cumulative percentages one can apply the formula:

$$\text{cum\% = (cum f)/N X 100}$$

III. Measures of Relative Standing. These are used to compare performances of an individual to that of a reference group.

A. Percentile point is the value on the measurement scale below which a specified percentage of scores in the distribution fall.

B. Formula:

$$\text{Percentile Point} = X_L + (i/f_i)(\text{cum } f_P - \text{cum } f_L)$$

C. Percentile rank of a score is the percentage of scores lower than the score in question.

D. Formula:

$$\text{Percentile rank} = \frac{\text{cum } f_L + (f_i/i)(X - X_L)}{N} \times 100$$

IV. Graphing Techniques

A. Graph characteristics.

    1. A graph has two axes, the vertical is called the ordinate and the horizontal is called the abscissa.

    2. Scores are shown on the abscissa; frequency is shown on the ordinate.

    3. Graphs are generally 3/4 as high as they are long.

B. Types of graphs for frequency distributions.

    1. Bar graphs are generally used for nominal or ordinal data. The abscissa shows categories and the ordinate frequency.

    2. Histograms are generally used for interval or ratio data. The abscissa shows the class interval for the data and the ordinate shows frequency. Vertical bars must touch each other.

    3. Frequency polygons are similar to histograms, except that instead of using bars, a point is plotted over the midpoint of each interval at a height corresponding to the frequency of the interval. In addition, the line is extended to meet the horizontal axis at the midpoint of the two end-adjacent intervals.

    4. Cumulative percentage polygons show the percentage of scores which fall below the upper real limit of the interval on the ordinate. The abscissa plots the upper limit of each class interval.

C. Shapes of frequency curves.

    1. Symmetrical distributions are those which when folded in half have the two sides of the curve correspond perfectly.

    2. Skewness. If a curve is not symmetrical, it is skewed.

        a. Positive skewness refers to curves where most of the scores occur at the lower values of the abscissa and tails off to the higher end.

        b. Negative skewness refers to curves where most of the scores occur at the higher values and the curve tails toward the lower end of the abscissa.

    3. Miscellaneous shapes. Curves are sometimes described in terms of their shape. Examples include U-shaped distributions, J-shaped distributions, rectangular, and bell-shaped.

D. Exploratory Data Analysis.

    1. Uses easy-to-construct diagrams to describe data.

2. Stem and leaf diagrams.

   a. Alternative to histogram, but unlike the histogram, perserves the orignial scores.

   b. Each score is represented by a stem and leaf. Stems are placed vertically down the page and leafs are placed to the right of the stems, horizontally across the page. Often, a vertical line is used to separate the stems from the leafs.

CONCEPT REVIEW

An unordered list of many scores is difficult to understand.

To report large amounts of data in an informative way, a

table is often presented. A (1) _____ presents the score

values and their frequency of occurrence. The scores are

listed in (2) _____ order, generally with the (3) _____ score

at the bottom of the table. To avoid having many values with

frequencies of zero, the individual scores are usually (4) _____

into (5) _____ intervals of equal width. This is called a

frequency distribution of (6) _____. The (7) _____ the interval

the more information is lost. If the interval is too (8) _____ then

too many zero frequencies occur. Generally the distribution is

divided into from (9) _____ to (10) _____ intervals.

To construct a frequency distribution, one employs the following

steps:

1. Find the (11) _____ of the scores. To determine the

range, one subtracts the (12) _____ value from the

(1) frequency distribution

(2) rank
(3) lowest

(4) grouped

(5) class

(6) grouped scores
(7) wider
(8) narrow

(9) 10
(10) 20

(11) range

(12) lowest

(13) _____ value in the distribution.

(13) highest

    2. Determine the (14) _____ of the class interval.

(14) width

The width of the class interval is given by the symbol

(15) _____. The width of the class interval is deter-

(15) i

mined by dividing the (16) _____ by the (17) _____ of class

(16) range
(17) number

intervals. If there is a decimal remainder, round to the

(18) _____ number of digits as the raw

(18) same

scores.

    3. When listing the intervals, begin with the (19) _____

(19) lowest

interval and proceed to the highest. First, one needs to deter-

mine the (20) _____ of the lowest interval. Two criteria exist in

(20) lower limit

determining the values of the lowest intervals. The interval must

begin with a value (21) _____ or (22) _____ the lowest

(21) equal to
(22) less than

score. Also, the lower limit should be evenly (23) _____

(23) divisible

by i.

    4. Next, the raw scores are (24) _____

(24) tallied

into the appropriate class intervals. This simply entails

entering a tally mark next to the interval containing the raw score.

    5. Count the tally marks and convert them into

(25) _____.

(25) frequencies

In addition to simply countinq up frequencies, it is often use-

ful to summarize them further. A (26) _____ frequency dis-

(26) relative

tribution indicates the proportion of the total number of scores

which occurred in each (27) _____. A (28) _____ fre-

(27) interval
(28) cumulative

quency distribution indicates the number of scores which

fell below the (29) _____ real limit of each interval. A

(29) upper

cumulative (30) _____ distribution indicates the percentage

(30) percentage

of scores which fell below the upper real limit of each interval.

To convert a frequency distribution into a relative frequency

distribution, the (31) _____ for each interval is (32) _____

(31) frequency
(32) divided

by the total number of scores. The cumulative frequency for

each interval is found by (33) _____ the frequency of that inter-

(33) adding

val to the sum of the frequencies of all the class intervals below.

The cumulative percentage for each interval is found simply

by converting cumulative (34) _____ to cumulative

(34) frequencies

(35) _____. The equation for this is:

(35) percentages

$$\text{cum\% = (cum f)/N X 100}$$

Cum% is the (36) _____. N equals the (37) _____ number

(36) cumulative percentage
(37) total

of scores. Cum f equals the cumulative (38) _____ of the class

(38) frequency

interval of interest. If there were 180 scores altogether and

the cumulative frequency of a given class interval equaled 40,

the cumulative percentage would be (39) _____. This would

(39) cum% = (40/180) x 100 = 22.22%

mean that (40) _____ percent of the scores fell below the

(40) 22.22

upper real limit of the interval in question.

Percentages are measures of relative standing. They are

used to (41) _____ the performance of an individual to that

41) compare

of a reference group. A (42) _____ is the value on the

(42) percentile or percentile

measurement scale below which point a specified percentage

of the scores in the distribution falls. The specified percentage

is called the (43) _____ of the value. For example, the 38th

percentile point is the value below which (44) _____ percent

of the scores in the distribution fall. This value has a percentile

(45) _____ of (46) _____ .

There ia a formula to compute the percentile point. The

formula is:

$$\text{Percentile Point} = X_L + (i/f_i)(\text{cum } f_P - \text{cum } f_L)$$

In order to avoid difficulties in understanding and applying

the equation, one needs to be certain that one understands what

the symbols mean. In this equation:

X_L = the value of the (47) _____ real limit of the interval

    (48) _____ the percentile point

cum f_P = the frequency of scores (49) _____ the percentile

    point.

cum f_L = the frequency of scores (50) _____ the

    (51) _____ real limit of the interval containing

    the percentile point.

f_i = the frequency of the interval (52) _____ the percentile

    point.

i = the (53) _____ of the (54) _____ .

Now is a good time to refresh the memory on the rules of

point

(43) percentile rank

(44) 38

(45) rank
(46) 38

(47) lower

(48) containing

(49) below

(50) below

(51) lower

(52) containing

(53) width
(54) interval

algebra as they apply to simplifying this equation. The first step

is to simplify the terms within (55) _____. The second step

is to (56) _____ the terms in parentheses together. Finally,

one adds this product to (57) _____ to obtain the

(58) _____.

It is often useful to know the percentile rank of a raw score.

The percentile rank of a score is the (59) _____ of scores

(60) _____ than the score in question. This process is

the (61) _____ situation than the process of calculating

the (62) _____.In this case we are given the (63) _____

and must calculate the (64) _____ of scores below it. In the

percentile point calculation we are given the (65) _____

and must calculate the (66) _____. The formula for the

percentile rank is as follows:

$$\text{Percentile rank} = \frac{\text{cum } f_L + (f_i/i)(X - X_L)}{N} \times 100$$

where

    cum f_L = the frequency of scores (67) _____ the (68) _____

        real limit of the interval containing the score X.

    X = the raw score whose (69) _____ is being determined.

    X_L = the (70) _____of the lower real limit of the

        interval containing (71) _____.

    i = (72) _____.

    f_i = the (73) _____ of the interval containing X.

(55) parentheses

(56) multiply

(57) X_L

(58) percentile point

(59) percentage

(60) lower

(61) opposite

(62) percentile point
(63) raw score
(64) percentage
(65) percentile

(66) scale value

(67) below
(68) lower

(69) percentile rank

(70) value

(71) X

(72) interval width

(73) frequency

N = the (74) _____ number of scores.

(74) total

The same rules of algebra apply to simplifying this equation as did for the previous one.

Graphs are often used to show frequency distributions. They present (75) _____ new information from the tables discussed above, but a picture can sometimes convey information more clearly. Graphs have two axes. The vertical axis is called the (76) _____. The horizontal axis is called the (77) _____. In graphing frequency distributions, the score (78) _____ are plotted along the (79) _____. It is customary to have the height of the graph be about (80) _____ of the length.

(75) no

(76) ordinate
(77) abscissa
(78) frequencies

(79) ordinate

(80) 3/4

One common way of plotting nominal or ordinal data is using the (81) _____ graph. A bar is drawn for each (82) _____ and the height of the bar represents (83) _____. In the case of (84) _____ data the order of arrangement of the groups along the horizontal axis does not matter. The bars do not touch each other to emphasize the lack of a (85) _____ relationship between the categories.

(81) bar

(82) category

(83) frequency
(84) nominal

(85) quantitative

A (86) _____ can be used to represent frequency distributions of interval or (87) _____ scales. In a histogram a bar is drawn for each (88) _____. Usually the (89) _____ of the class intervals are plotted along the (90) _____. The height of the bar corresponds to the (91) _____ of the class interval. Since the data are

(86) histogram

(87) ratio

(88) class interval

(89) midpoints

(90) abscissa

(91) frequency

(92) _____, the bars touch each other.   (92) continuous

A frequency (93) _____ can also be used to represent   (93) polygon

interval or ratio data. Instead of using bars, a point is plotted

over the (94) _____ of each class interval at a height   (94) midpoint

corresponding to the (95) _____ of the interval. At the ends of   (95) frequency

the distribution, the lines joining the points are extended to meet

the (96) _____ axis to form a polygon.   (96) horizontal

The cumulative percentage of a distribution can easily

be calculated and, therefore, graphed. When graphed this is

called a (97) _____ curve. In this graph the vertical   (97) cumulative
                                                                 percentage
 axis is scaled in (98) _____ units. On the horizontal   (98) cumulative
                                                                 percentage
axis the upper real limit of the class interval is plotted.

One can read both (99) _____ and (100) _____ from a   (99) percentiles
                                                            (100) percentile rank
cumulative percentage curve. The cumulative percentage

curve is also called an (101) _____, meaning   (101) ogive

S-shape.

Frequency curves can be described in terms of their

shapes. They are generally classified as to whether they are

(102) _____ or (103) _____. A curve is (104) _____   (102) symmetrical
                                                              (103) skewed
if when folded in half the two sides coincide. Distributions   (104) symmetrical

which are not symmetrical are (105) _____. A positively   (105) skewed

skewed distribution has most of the scores at the

(106) _____ values of the distribution and tails off to the   (106) lower

(107) _____ values. A negatively skewed distribution   (107) higher

has most of the scores at the (108) _____ values of

the distribution and tails off toward the (109) _____ end of

the horizontal axis.

    Frequency distributions are often referred to according to

their _____; e.g., bell-shaped, U-shaped, J-shaped, or

rectangular, to just name a few.

(108) higher

(109) lower

(110) shape

EXERCISES

1. The first step in making a frequency distribution is to rank order the scores. Generally this is done by ranking scores from lowest to highest. Rank order the following scores:

30, 20, 21, 25, 5, 18, 18, 16, 10

2. What is the range of the distribution in problem 1?

3. Assume the following are scores in a 100 point achievement test:

58	70	68	59	62
47	90	57	45	68
93	83	45	48	55
80	61	80	70	51
35	58	36	63	71
60	41	65	84	42
40	87	75	81	80
61	72	76	73	52
63	71	94	66	
68	69	60	61	

a. What is the range of the distribution?

b. Since there is no definitive rule for determining how many class intervals to divide data into, one generally plays with the data or is told how many intervals to use. If you were told to group the above data into approximately 12 intervals of equal width, how wide would the class intervals be? Using the value just calculated for i, what would be the lower apparent limit of the lowest class interval?

c. Using 12 intervals construct a table which shows the frequency distribution of grouped scores, the corresponding cumulative frequency distribution, and the corresponding cumulative percentage distribution.

d. Plot the cumulative percentage curve for this data.

4. You have just been hired as a statistician by the federal government. Your first assignment is to prepare a graphic representation of the number of types of animals in a certain national forest. You are told there are 525 birds, 100 beavers, 150 bears, 300 deer, 250 elk, 400 squirrels, 600 rabbits, and 200 geese. Prepare a graph to show these data.

5. Sketch a cumulative frequency distribution for the following frequency distribution.

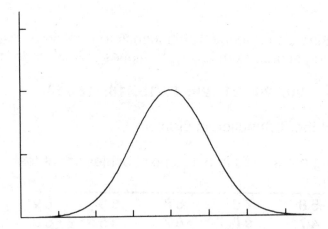

6. Consider the frequency distribution shown in the table below.

class interval	f	cum f	cum %
45-49	2	128	100.00
40-44	6	126	98.44
35-39	15	120	93.75
30-34	24	105	82.03
25-29	30	81	63.28
20-24	22	51	39.84
15-19	16	29	22.66
10-14	8	13	10.16
5-9	4	5	3.91
0-4	1	1	0.78

a. What is the percentile point for the 25th percentile (P_{25})?
b. For the 50th percentile (P_{50})?
c. For the 75th percentile (P_{75})?

7. Again consider the data from problem 6.

    a. What percentile rank would a score of 24 have?
    b. What percentile rank would a score of 32 have?
    c. What percentile rank would a score of 35 have?

8. Plot a cumulative percentage distribution for the data in problem 6.

9. What is wrong with the following frequency distribution?

class interval	f
80-99	51
60-79	59
40-59	112
20-39	49
0-19	50

10. Assume the following are weights of a sample of junior high school students.

98	133	108	106	104
124	110	137	119	110
120	130	112	129	126
112	111	120	103	124
126	119	130	115	118
102	121	125	118	122
116	139	134	123	114
106	112	108	116	120

    a. Construct a frequency distribution for the data with approximately 10 intervals.
    b. What is the value for P_{50}?
    c. What is the value for P_{90}?
    d. What is the percentile rank of a weight of 133 pounds?
    e. What is the percentile rank of 105 pounds.

11. Construct a stem and leaf diagram for the frequency distribution shown in problem 10.

Answers: 1. 5, 10, 16, 18, 18, 20, 21, 25, 30   **2.** 25   **3a.** 59   **3b.** 5, 35   **3c.**

class interval	f	cum f	cum %
90-94	3	48	100.00
85-89	1	45	93.75
80-84	6	44	91.67
75-79	2	38	79.17
70-74	6	36	75.00
65-69	6	30	62.50
60-64	8	24	50.00
55-59	5	16	33.33
50-54	2	11	22.92
45-49	4	9	18.75
40-44	3	5	10.42
35-39	2	2	4.17

4. You should have drawn a bar graph since we are showing nominal data on the X axis
5. You should have drawn an ogive   **6a.** 20.18   **6b.** 26.67   **6c.** 32.62   **7a.** 38.12
7b. 72.66   **7c.** 83.20   **9.** class intervals are too wide   **10a.**

class interval	f
136-139	2
132-135	2
128-131	3
124-127	5
120-123	6
116-119	6
112-115	5
108-111	5
104-107	3
100-103	2
96-99	1

10b. P_{50} is 118.17   **10c.** 131.5   **10d.** 91.88   **10e.** 10.31.

11.

9	8
10	234
10	6688
11	0012224
11	5668899
12	00012344
12	5669
13	0034
13	79

TRUE-FALSE QUESTIONS

T F   1. When constructing frequency distributions there must be 12 class intervals.

T F   2. In a frequency distribution the more intervals the better, regardless of whether some intervals have zero frequency.

T F   3. One reason for constructing frequency distributions is to be able to visualize the shape of the distribution.

T F   4. When constructing the frequency distribution it is customary to show the real limits of the class intervals in the table.

T F   5. A cumulative frequency distribution indicates the number of scores which fell below the upper real limit of each interval.

T F   6. A relative frequency distribution indicates the total number of scores which occurred in each interval.

T F   7. The vertical axis, i.e., the Y axis of a graph, is called the abscissa.

T F   8. Bar graphs are generally used for nominal or ordinal data and histograms are generally used for interval or ratio data.

T F   9. A "U" shaped distribution is an example of a symmetrical distribution.

T F 10. In a cumulative percentage curve, percentage is shown on the abscissa.

T F 11. To determine the width of the class interval, i, divide the range by the number of class intervals.

T F 12. The lower limit in the lowest class interval should equal i.

T F 13. Stem and leaf diagrams are like histograms, except that they lose the original scores.

Answers: 1. F   **2.** F   **3.** T   **4.** F   **5.** T   **6.** F   **7.** F   **8.** T   **9.** T   **10.** F **11.** T   **12.** F   **13.** F.

SELF-QUIZ

1. The purpose of a frequency distribution is to _____.

   a. present scores and their frequency of occurrence
   b. present data in a more meaningful way than single scores
   c. provide more information than a graph
   d. all of the above
   e. a and b

2. The true limits of 7.0 are _____.

   a. 6.5-7.5
   b. 6.0-8.0
   c. 7.0-7.1
   d. 6.95-7.05

3. When individual scores are combined into groups, _____.

   a. information is lost
   b. data is added
   c. a meaningful visual display can result depending on the interval width
   d. a and c

4. The range of a set of scores with a maximum value of 92 and a minimum value of 26 is _____.

   a. 65
   b. 66
   c. 67
   d. 92

5. If the range of a distribution were 89 and the data were reported as whole numbers, what would the width of the class interval be if one chose to group the distribution into approximately 14 class intervals?

   a. 14
   b. 89
   c. 5
   d. 6
   e. 7

6. If i = 7, and the minimum value of a distribution of scores was 8, what would the lowest class interval be?

   a. 0-7
   b. 7-14
   c. 8-16
   d. 8-15
   e. 7-13

7. What indicates the proportion of the total number of scores which occurred in each interval?

   a. relative frequency distribution
   b. cumulative frequency distribution
   c. cumulative percentage distribution
   d. none of the above

8. What indicates the number of scores which fell below the upper real limit of each interval?

   a. relative frequency distribution
   b. cumulative frequency distribution
   c. cumulative percentage distribution
   d. none of the above

9. In graphing frequency distributions, _____ is usually plotted on the abscissa.

   a. frequency
   b. class width
   c. the score value
   d. interval width

10. When constructing bar graphs, the bars do not touch each other because _____.

   a. it looks nicer
   b. it emphasizes the lack of quantitative relationship between the categories
   c. it is traditional
   d. none of the above

11. In a frequency polygon the points are plotted over _____ at a height corresponding to the frequency of the interval.

   a. the midpoint of each interval
   b. the lower real limit
   c. the upper real limit
   d. none of the above

12. Which of the following is (are) not a symmetrical distribution?

    a. a bell-shaped curve
    b. a J-shaped curve
    c. a rectangular curve
    d. an inverted U-shaped curve
    e. a, b and d

13. A distribution which has a predominance of scores at the lower values of the distribution and which tails off at the higher end is _____.

    a. positively skewed
    b. negatively skewed
    c. normally distributed
    d. symmetrical

For questions 14 - 22 refer to the following table.

class interval	f	cum f	cum %	relative %
90-99	2	139	100.00	1.44
80-89	7	137	98.56	5.04
70-79	15	130	93.53	10.79
60-69	21	115	82.73	15.11
50-59	37	94	67.63	26.62
40-49	26	57	41.01	18.71
30-39	19	31	22.30	13.67
20-29	6	12	8.63	4.32
10-19	4	6	4.32	2.88
0-9	2	2	1.44	1.44

14. How many occurrences are there for the interval 60-69?

15. How many occurrences fall below the upper real limit of the interval 70-79?

16. What percentage of cases fall within the class interval containing the most cases?

17. N equals _____.

18. What is the cumulative percentage below the lower real limit of the interval 90-99?

19. What is the value of i?

20. The cumulative frequency of 115 indicates that 115 scores fall below _____.

    a. 115
    b. 60
    c. 69
    d. 69.5
    e. 60.5
    f. 64.5

21. The percentile rank of a score of 41 equals _____.

22. The 50th percentile point equals _____.

Answers: 1. e **2.** d **3.** d **4.** b **5.** d **6.** e **7.** a **8.** b **9.** c **10.** b **11.** a **12.** b **13.** a **14.** 21 **15.** 130 **16.** 26.62 **17.** 139 **18.** 98.56 **19.** 10 **20.** d **21.** 25.11% **22.** 52.88.

4 | MEASURES OF CENTRAL TENDENCY AND VARIABILITY

CHAPTER OUTLINE

I. **Measures of Central Tendency**. There are three commonly used measures of central tendency.

    A. <u>Arithmetic mean</u>. Symbolized by "X" (read $\overline{X}$ bar), or "μ" (read mu). This is simply the average of a set of raw scores. When referring to an entire population, the symbol used is μ. $\overline{X}$ is used for sample data.

        1. Equation for calculating the mean of sample raw scores

$$\overline{X} = (\Sigma\ X_i)/N$$

        This equation simply tells us to add all of the scores and divide by the total number of scores.

        a. The mean is sensitive to the exact value of all the scores in the distribution.

        b. $\Sigma\ (X - \overline{X}) = 0$. The sum of the deviations about the mean equals zero.

        c. The mean is very sensitive to extreme scores when the scores are not balanced at both ends of the distribution.

43

d. $\Sigma (X - \overline{X})^2$ is a minimum. The sum of the squared deviations of all the scores about their mean is a minimum.

e. Under most circumstances, of the measures used for central tendency, the mean is least subject to sampling variation.

3. Overall mean equation.

$$\overline{X}_{overall} = \frac{\text{sum of all scores}}{N} = \frac{n_1 \overline{X}_1 + n_2 \overline{X}_2 + \cdots + n_k \overline{X}_k}{n_1 + n_2 + \cdots + n_k}$$

which signifies that the overall mean is equal to the sum of the mean of each group times the number of scores in the group, divided by the sum of the number of scores in each group.

B. <u>The median (Mdn)</u> This is the value below which 50% of the scores fall. It is the same thing as P_{50}.

1. Equation for grouped scores:

$$\text{Mdn} = X_L + (i/f_i)(\text{cum } f_P - \text{cum } f_L)$$

2. Calculation with raw scores. The median is the centermost score, if the number of scores is odd. If the number is even, the median is taken as the average of the two centermost scores. (Note: scores are rank ordered first.)

3. Properties of the median:

a. The median is less sensitive than the mean to unbalanced extreme scores.

b. Under usual circumstances, the median is more subject to sampling variability than the mean, but less subject to sampling variability than the mode.

C. <u>The mode</u>. This is simply the most frequent score in the distribution. It is calculated by inspection.

D. <u>Symmetry</u>. In a unimodal (one mode), symmetrical distribution, the mean, median and mode are all equal. In a positively skewed distribution, the mean will be larger than the median. In a negatively skewed distribution, the mean will be lower than the median.

II. Measures of Variability. These measures quantify the extent of dispersion in a distribution.

    A. <u>Range</u>. The range is the difference between the highest and lowest score in the distribution. The range measures only the spread of the two extreme scores.

    B. <u>Deviation scores</u>. These tell how far away the raw score is from the mean of its distribution. It is symbolized by $(X - \bar{X})$ or $(X - \mu)$.

$$(X - \bar{X}) \qquad \text{for sample data}$$

$$(X - \mu) \qquad \text{for population data}$$

    C. <u>The standard deviation</u>. This is a measure of an average deviation of raw scores about the mean. It is symbolized by "s" for the sample standard deviation, and by "σ" (small sigma) for the population standard deviation.

        1. Conceptual formula:

$$s = \sqrt{\frac{\Sigma (X - \bar{X})^2}{N-1}} = \sqrt{\frac{SS}{N-1}}$$

        2. Computational formula:

$$s = \sqrt{\frac{\Sigma X^2 - \frac{(\Sigma X)^2}{N}}{N-1}}$$

because $\Sigma (X - \bar{X})^2 = \Sigma X^2 - (\Sigma X)^2 / N$

    3. For the population standard deviation, the formula is just a little different. The denominator of the formula is N, instead of N - 1.

        a. Conceptual formula:

$$\sigma = \sqrt{\frac{\Sigma (X - \mu)^2}{N}} = \sqrt{\frac{SS_{pop}}{N}}$$

b. Computational formula:

$$\sigma = \sqrt{\frac{\sum x^2 - \frac{(\sum x)^2}{N}}{N}}$$

4. Properties of the standard deviation.

   a. The standard deviation gives a measure of dispersion relative to the mean.

   b. The standard deviation is sensitive to each score in the distribution.

   c. It is stable with regard to sampling fluctuations.

D. <u>The variance</u>. This is simply the square of the standard deviation. The variance is symbolized as "s2" for the sample, and "σ2" for the population. This is read as "the variance" or "s squared."

   1. Conceptual formula:

$$s^2 = \frac{\sum (X - \overline{X})^2}{N-1} = \frac{SS}{N-1} \qquad \text{for a sample}$$

$$\sigma^2 = \frac{\sum (X - \mu)^2}{N} = \frac{SS_{pop}}{N} \qquad \text{for a population}$$

   2. Computational formula:

$$s^2 = \frac{\sum x^2 - \frac{(\sum x)^2}{N}}{N-1} \qquad \text{for a sample}$$

$$\sigma^2 = \frac{\sum x^2 - \frac{(\sum x)^2}{N}}{N} \qquad \text{for a population}$$

CONCEPT REVIEW

Organizing and presenting data is often a problem for

scientists. We have just discussed frequency distributions.

Frequency distributions do not allow (1) _____.

statements about the distributions. Nor do they allow

quantitative (2) _____ to be made between two distributions.

Two characteristics of distributions often described are the

(3) _____ tendency and (4) _____.

Central tendency refers to how scores tend to (5) _____

around a central point in the distribution. Variability refers to the

extent to which scores are (6) _____ or spread out.

There are three common measures of central tendency.

They are the (7) _____, (8) _____ and (9) _____. The

arithmetic mean is what we commonly compute when we com-

pute the (10) _____ of a set of scores. The mean is the

(11) _____ of the scores divided by the (12) _____ of scores.

We let the symbol (13) _____ stand for the mean of sample

scores. The equation for the mean of sample scores is $\overline{X}$ =

(14) _____. For a population mean we use the same equa-

tion except we let (15) _____ stand for the mean instead

of $\overline{X}$. The symbol μ is the Greek letter mu, pronounced "mew."

Assume that you have observed the following scores

on a test: 23, 30, 17, 20, and 25. N equals (16) _____. ΣX_i

equals (17) _____, and the mean equals (18) _____. If you

had the ages of all people at your school and compiled the

average age for the school, you would be computing the

(1) quantitative

(2) comparisons

(3) central
(4) variability
(5) cluster

(6) dispersed

(7) mean
(8) median
(9) mode

(10) average

(11) sum
(12) number
(13) $\overline{X}$

(14) $\Sigma X_i/N$

(15) μ

(16) 5

(17) 115
(18) 23

population mean, so you would use the symbol (19) _____ .   (19) μ

The mean possesses several interesting properties. The

first property of the mean is that it is sensitive to the (20) _____   (20) exact

value of all of the scores in the distribution. This means if one

changes the value of any of the scores, this will cause a

(21) _____ in the value of the mean, If one changes the   (21) change

value of more than one score, the changes may cancel each

other out and the mean could stay the (22) _____ . But in   (22) same

general if you change (23) _____ the mean will (24) _____ .   (23) scores
(24) change

Another important property of the mean is that the sum

of the deviations about the mean equals (25) _____ .   (25) zero

Written algebraically, this is (26) _____ = 0. This property says   (26) $\Sigma\,(X - \overline{X})$

that if the (27) _____ is subtracted from each   (27) mean

(28) _____ , the sum of the (29) _____   (28) score
(29) differences

will equal zero. $(X - \overline{X})$ is referred to as the

(30) _____ .   (30) deviation score

A third property of the mean is that it is very sensitive to

(31) _____ scores when the scores are not (32) _____ at   (31) extreme
(32) balanced

both ends of the distribution. For example, the mean of the

numbers 8, 8, 9, 10, and 11 is (33) _____ . The mean of   (33) 9.2

8, 8, 9, 10, 11 and 100 is (34) _____ . We can see how   (34) 24.33

adding one extreme score more than doubles the mean.

A fourth property of the mean is that the (35) _____ of the   (35) sum

(36) _____ deviations of all the scores about their mean   (36) squared

is a (37) _____.. The algebraic symbols designating the sum

of the squared deviations of all scores about their mean is

(38) _____.

The final property of the mean we will discuss is that under

most circumstances, of the measures of central tendency, the

mean is the (39) _____ subject to (40) _____ variation.

This means that if we drew many (41) _____ samples from a

(42) _____, the means would probably be different for each

(43) _____ but the differences would be less for the mean than

for the (44) _____ or (45) _____.

The overall mean for several groups of scores is sometimes

useful to calculate. For k groups the overall mean is found by:

$$\overline{X}_{overall} = \frac{\textbf{sum of all scores}}{N}$$
$$= \frac{n_1\overline{X}_1 + n_2\overline{X}_2 + \cdots + n_k\overline{X}_k}{n_1 + n_2 + \cdots + n_k}$$

This is true because $\overline{X}_1 = \Sigma X_1/n_1$ can be transformed by

multiplying the above equation by (46) _____. This gives

ΣX_1 (first group) = (47) _____. This means that the overall

mean is equal to the sum of the (48) _____ of each group

times the number of scores in the group, (49) _____

by the sum of the number of scores in each group. If you are

told that $\overline{X}_1 = 42$ and $n_1 = 20$, then ΣX_1 must equal

(50) _____. Suppose you are told that the mean grade in

(37) minimum

(38) $\Sigma (X - \overline{X})^2$

(39) least
(40) sampling
(41) random

(42) population

(43) sample

(44) median
(45) mode

(46) n_1

(47) $n_1\overline{X}_1$

(48) mean

(49) divided

(50) 840

one class was 78 for 250 students and 83 for 100 students in a second class. The overall mean for the two classes equals:

$$\overline{X}_{overall} = \frac{(51)\ \underline{\quad} \times 250 + (52)\ \underline{\quad} \times 100}{(53)\ \underline{\quad} + (54)\ \underline{\quad}}$$

$$= (55)\ \underline{\quad}.$$

(51) 78
(52) 83
(53) 250
(54) 100
(55) 79.43

The median is the value below which (56) _____ % of

the scores fall. It is the same thing as the scale value at

(57) _____. For grouped data to calculate the (58) _____

all one must do is calculate (59) _____. To calculate the

median for raw scores you must first (60) _____ the scores.

If the number of scores is odd, the median is the (61) _____

score. If the number of scores is even, the median is taken as the

(62) _____ of the two centermost scores. So for the scores

1, 2, 3, 4, 5, and 6, the median is (63) _____.

Like the mean, the median has a couple of properties which

are useful to know. First, the median is (64) _____ sensitive

than the mean to unbalanced scores. For example, the mean

of 1, 3 and 5 is (65) _____, and the median of 1, 3 and 100

is (66) _____. Thesecond property of the median is that it is

(67) _____ subject to sampling variability than the mean, but

(68) _____ subject to sampling variability than the

(69) _____.

(56) 50

(57) P_{50} or 50th percentile
(58) median
(59) P_{50}

(60) rank order

(61) centermost

(62) average

(63) $(3 + 4)/2 = 3.5$

(64) less

(65) 3

(66) 3

(67) more

(68) less

(69) mode

The mode is the last measure of central tendency we will

discuss. The mode is the (70) _____ score in the distribution.   (70) most frequent

Determining the mode is simply done by (71) _____. In the   (71) inspection

scores 90, 92, 93, 96, 100, 100 and 100, the mode is

(72) _____. For the scores 1, 1, 1, 1, 2, 3, 9, 9, 9, and   (72) 100

9, there are (73) _____ modes. The values for the modes   (73) two

are (74) _____ and (75) _____.This is   (74) 1
     (75) 9
called a (76) _____distribution. In a frequency poly-   (76) bimodal

gon the mode is the (77) _____ point on the graph. If a   (77) highest

distribution is (78) _____ and (79) _____ the mean,   (78) unimodal
     (79) symmetrical
median and mode are all equal. An example of an unimodal,

symmetrical distribution is the (80) _____ curve. When the   (80) bell-shaped

distribution is skewed, the median and mean will not be equal.

The (81) _____ is sensitive to extreme scores. When the   (81) mean

distribution is negatively skewed the mean will be (82) _____   (82) less

than the median. When the distribution is (83) _____ the   (83) positively
     skewed
mean will be greater than the median.

Measures of Variability

Variability has to do with how far scores are (84) _____.   (84) spread apart

Therefore, measures of variability (85) _____ the extent   (85) quantify

of dispersion. There are three measures of variability; the

(86) _____, (87) _____, and the (88) _____.   (86) range
     (87) standard
     deviation
     (88) variance
We have already looked at the range. The range   (89) highest

is the difference between the (89) _____and   (89) highest

(90) _____scores in the distribution. So the range of the   (90) lowest

scores 18, 20, 27, 30, 32 and 41 is (91) _____ - (92) _____   (91) 41
                                                                    (92) 18
= (93) _____. The range is easy to calculate but of limited   (93) 23

use because it measures the spread of the two (94) _____   (94) extreme

scores but not the spread of the scores in between.

One might be more interested in how far away scores are

from the mean. A (95) _____ score tells how far away the   (95) deviation

raw score (X) is from the mean ($\overline{X}$). Mathematically this

is stated (96) _____.   (96) $X - \overline{X}$

The (97) _____ deviation uses the (98) _____ scores of   (97) standard
                                                                     (98) deviation
the data. The standard deviation is symbolized by an

(99) _____ for sample data and the Greek letter   (99) s

(100) _____ for population data. The formula for the   (100) σ

standard deviation of a (101) _____ is   (101) population

$$\sigma = \sqrt{\frac{\Sigma (X - \mu)^2}{N}}$$

In words, this means subtract the value of the (102) _____   (102) mean

from each raw score, square the result, sum each

of the resulting (103) _____ scores, divide this sum   (103) squared
                                                                          deviation
by (104) _____, and then take the (105) _____. The   (104) N
                                                                     (105) square root
reason we square the (X - μ) term is because of one of the

properties of the mean. Since $\Sigma (X - \mu)$ equals (106) _____, we   (106) zero

can (107) _____ each (X - μ) term so the positive and   (107) square

negative differences don't cancel. To counteract squaring

each term, we unsquare everything at the end by taking the

(108) _____ of $\sum (X - \mu)^2/N$. Thus the final equation for the

standard deviation of the population is

$$\sigma = \sqrt{\frac{\sum (X - \mu)^2}{N}}$$

The equation for the standard deviation of a (109) _____

uses (110) _____ in the denominator rather than

(111) _____ to give a more accurate estmate of the population

standard deviation. So the conceptual formula for the sample

standard deviation is

(109) sample

(110) N - 1
(111) N

$$s = \sqrt{\frac{\sum (X - \overline{X})^2}{N - 1}} = \sqrt{\frac{SS}{N - 1}}$$

The numerator of the formula, $\sum (X - \overline{X})^2$, is also symbolized

by SS. SS stands for (112) _____.

(112) sum of squares

Let's practice calculating "s" on the following sample

data: 21, 24, 27, 30, and 33. Fill in the following table:

X	X - $\overline{X}$	(X - $\overline{X}$)^2	
21	21 - 27	$(-6)^2$ = 36	
24	24 - 27	$(-3)^2$ = 9	
27	27 - 27	$(0)^2$ = 0	
30	30 - 27	$(3)^2$ = 9	
33 (113) _____	(114) ___ = ___ (115)		

$\sum X$ = (116) _____; $\sum (X - \overline{X})^2$ = (117) _____;

N = (118) _____; $\overline{X} = \sum X/N$ = (119) _____ = 27.0

(113) 33 - 27
(114) $(6)^2$
(115) 36
(116) 135
(117) 90
(118) 5
(119) 135/5

This (120) _____ formula is easy to use when the mean is

(120) conceptual

(121) _____ fractional. However, it can be shown that

$\Sigma (X - \overline{X})^2 =$ (122) _____ which allows a new formula

for the computation of s   This computational formula is:

$$s = \sqrt{\dfrac{\Sigma X^2 - \dfrac{(\Sigma X)^2}{N}}{N - 1}}$$

It is the preferred formula to use when the mean is (123) _____

Applying the computational formula to the same data, we obtain:

X	X^2
2 1	441
2 4	576
2 7	(124) ___
3 0	900
3 3	1089

$\Sigma X = 135$; $\Sigma X^2 =$ (125) _____; $(\Sigma X)^2 =$ (126) _____;

$$s = \sqrt{\dfrac{\Sigma X^2 - \dfrac{(\Sigma X)^2}{N}}{N - 1}} = \sqrt{\dfrac{(127) \underline{\quad} - \dfrac{(128) \underline{\quad}}{5}}{5 - 1}} = 4.74$$

The standard deviation has certain characteristics. First, the standard deviation gives us a measure of dispersion relative to the (129) _____. Second, like the mean, the standard deviation is sensitive to (130) _____ in the distribution. Finally, like the mean, the standard deviation is (131) _____ with regard to sampling fluctuation.

(121) not

(122) $\Sigma X^2 - (\Sigma X)^2/N$

(123) fractional.

(124) 729

(125) 3735
(126) 18,225

(127) 3735

(128) 18,225

(129) mean

(130) each score

(131) stable

The variance is the last measure of dispersion to be dis-

cussed. The variance of a set of scores is just the

(132) _____ of the standard deviation. For a population we

designate the variance as (133) _____. For a sample

we again use (134) _____ as the denominator, because N

gives too small an estimate for σ^2 when using sample data.

Almost always, one calculates the (135) _____ before the

standard deviation since one has to take the (136) _____ of

the variance to get the (137) _____.

(132) square

(133) σ^2

(134) N - 1

(135) variance

(136) square root

(137) standard deviation

EXERCISES

1. The following IQ data has been obtained for 11 incoming graduate students:

   120, 130, 132, 139, 160, 115, 120, 142, 148, 120, 141

   Find each of the following statistics:

   a. mean
   b. median
   c. mode
   d. range

2. As a statistician (use your imagination), you are asked to recommend which measure of central tendency to use to characterize the following scores:

   0.6, 0.2, 0.1, 0.2, 0.2, 0.4, 0.3, 0.7, 0.8, 0.1, 0.0, 0.5, 22.5, 0.4.

   What is your recommendation and why?

3. An instructor decides to weight the scores on three tests differently. The first test is weighted only half as much as the second test. The third test is the final and it is weighted twice as much as the second test. The mean grade on the first test was 68, the mean of the second test was 75, and the mean on the third test was 80. What is

the mean grade for all three tests, assuming the same number of students took all the exams?

4. What is the value of $\sum X$ if 900 students had a mean of 78.2 on a test?

5. Given the following sample scores on an aptitude test?

30, 20, 15, 19, 16

   a. What is the mean of the scores?
   b. What is the value of $\sum (X - \bar{X})$?
   c. What is the standard deviation using the conceptual formula?
   d. What is the value of $\sum (X - \bar{X})^2$?
   e. What is the standard deviation using the computational formula?
   f. What is the value of $\sum (X - \bar{X})^2$ using the computational formula?
   g. What is the variance of these data?

6. Assume that the standard deviation of one distribution equals 15 and the standard deviation of another distribution is equal to 23. Which of the following distributions is more likely to represent the case where s = 15 if the other distribution's standard deviation equals 23?

a. 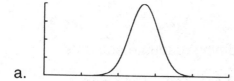   b.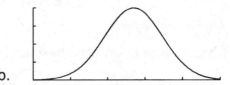

7. Consider the following two sets of scores.

X: 80, 70, 55, 63, 72
Y: 50, 52, 59, 60

   a. Which of these sets of scores was more likely drawn from a population with the greater variance?
   b. Why?

8. Consider two more sets of scores. Notice that the Y scores are just the X scores with a constant value of 10 added to each score. What effect does adding a constant have on the value of the standard deviation? Do you see why?

X: 1, 3, 5, 7, 9
Y: 11, 13, 15, 17, 19

9. A car company wanted to know how many miles per gallon an experimental car got after a series of tests. The following data were obtained.

Miles per gallon: 31.2, 28.6, 36.4, 37.3, 30.1, 29.0, 32.7, 31.9

a.   What is the mean number of miles per gallon obtained in this sample?
b.   What is the value of $\sum X$ and $\sum X^2$?
c.   What is the standard deviation of this sample?
d.   What is the variance of the sample?

10. Consider the same values as above except now each value is multipled by 2. The results are:

Miles per gallon: 62.4, 57.2, 72.8, 74.6, 60.2, 58.0, 65.4, 63.8

a.   What is the mean of this data?
b.   What is the standard deviation?
c.   What is the variance of the sample?

11. Repeat the above analysis on the original data in problem 9; this time multiply each original value by 4. The resulting data are:

Miles per gallon: 124.8, 114.4, 145.6, 149.2, 120.4, 116.0, 130.8, 127.6

a.   What is the mean now?
b.   What is the standard deviation?
c.   What is the variance?
d.   What is the relationship between the constant used to multiply the original data and the resulting mean, standard deviation and variance?

12. A statistics class taught by Dr. X had 30 students who obtained an average test score of 82.5. A class taught by Dr. Why had 20 students who ended up with an average score of 79.6. Dr. Zee (you guessed it) had 23 students who obtained an average score of 96.0. What was the average score for all the students combined?

Answers: 1a. 133.36   **1b.** 132   **1c.** 120   **1d.** 45   **2.** median. Mean would be too influenced by value 22.5. Since there's no impressive mode in the data, the median would appear to be the best choice.   **3.** 76.86   **4.** 70,380   **5a.** 20.0   **5b.** 0   **5c.** 5.96 **5d.** 142   **5e.** 5.96   **5f.** 142   **5g.** 35.5   **6.** a   **7a.** X   **7b.** The standard deviation of the X scores is 9.46 while the standard deviation of the Y scores is only 4.99 suggesting that Y may have been drawn from a population with less variance than X.   **8.** no effect on s because X increases by the same constant that each individual score does so the value of $\sum (X - \overline{X})^2$ does not change   **9a.** 32.15   **9b.** $\sum X = 257.2$; $\sum X^2 =$ 8,341.56   **9c.** 3.22   **9d.** 10.37   **10a.** 64.3   **10b.** 6.44   **11a.** 128.6   **11b.** 12.88 **11c.** 165.90   **lld.** The new mean is equal to the original mean times the constant. The same is true of the new standard deviation. The new variance is equal to the original variance times the square of the constant (within our rounding errors).   **12.** 85.96.

TRUE-FALSE QUESTIONS

T F 1. Central tendency and variability are two terms which mean the same thing.

T F 2. The mean is the same as the arithmetic average.

T F 3. By changing the value of one score in a sample this will necessarily cause the mean to change.

T F 4. By changing the value of one score in a sample this will necessarily cause the median to change.

T F 5. By changing the value of one score in a sample this will necessarily cause the mode to change.

T F 6. The mean is the most sensitive measure of central tendency to extreme scores.

T F 7. $\sum (X - \overline{X}) = 0$

T F 8. $\sum (X - \overline{X})^2 = 0$

T F 9. Of the measures of central tendency, the mean is the least subject to sampling variations.

T F 10. If a group of 100 people has an average IQ of 107 and a group of 50 people has an average IQ of 101, the average IQ of all the people is 105.

T F 11. It is possible to have more than one mode in a sample of scores.

T F 12. The formula for the population standard deviation is $\sqrt{\sum (X - \overline{X})^2/(N - 1)}$

T F 13. Measures of variability quantify the extent of dispersion in a set of scores.

T F 14. The variance is the square of the standard deviation.

T F 15. The symbol SS stands for sum of squares.

T F 16. If one knows $\sum X$, N, and $\sum X^2$, one can calculate both the variance and the standard deviation. It is not necessary to know the mean itself.

T F 17. If one uses N instead of N - 1 in the denominator of the equation for the sample standard deviation, one will underestimate the population standard deviation.

T F 18. The equation used for sample standard deviation is an estimate of the population standard deviation.

Answers: 1. F   **2.** T   **3.** T   **4.** F   **5.** F   **6.** T   **7.** T   **8.** F   **9.** T   **10.** T   **11.** T   **12.** F   **13.** T   **14.** T   **15.** T   **16.** T   **17.** T   **18.** T

SELF-QUIZ

1. The mean is _____ sensitive to extreme scores than the median.

   a. more
   b. less
   c. equally
   d. can't say without the scores

2. In a negatively skewed distribution, the mean will generally be _____ the median.

   a. greater than
   b. less than
   c. equal to
   d. can't tell from this information

3. $\sum (X - \overline{X})^2$ is called _____ .

   a. standard deviation
   b. mean
   c. variance
   d. sum of squares
   e. confusing

4. If 5 points are added to each score in a distribution, which of the following will happen?

   a. the mean will be unchanged
   b. the standard deviation will be unchanged
   c. the mode will be unchanged
   d. the mean, median and mode will be unchanged

5. $\sum (X - \overline{X})$ equals _____ .

   a. the sum of squares
   b. 0
   c. 1
   d. cannot be determined

6. $\Sigma (X - \overline{X})^2$ equals _____ .

    a. 0
    b. 1
    c. the variance
    d. cannot be determined from the information provided

7. Which of the following is not a measure of central tendency?

    a. mode
    b. median
    c. mean
    d. standard deviation

8. $\Sigma (X - \overline{X})^2$ equals _____ .

    a. ΣX^2
    b. $\Sigma X^2 - (\Sigma X)^2/N$
    c. a and b
    d. cannot be determined from information provided
    e. sum of squares
    f. b and e

9. If s = 12 for a set of 200 scores and 5 points were subtracted from each of the scores, the new value of the standard deviation would be _____ .

    a. 12/5
    b. $\sqrt{12/5}$
    c. 12 - 5
    d. 12

10. Assume one has a set of 35 scores with $\overline{X}$ = 100 and s = 9. If 5 scores, each equal to 100 were added to the distribution, what effect would this have on $\overline{X}$?

    a. increase $\overline{X}$
    b. decrease $\overline{X}$
    c. $\overline{X}$ would remain unchanged
    d. cannot tell

11. Considering again the information in problem 10, what would be the effect of adding those same new scores on the value of s?

    a. increase s
    b. decrease s
    c. s would remain unchanged
    d. cannot be determined from the information given

12. What is the appropriate symbol for the formula

$$\sqrt{\frac{\Sigma (X - \mu)^2}{N}}$$

    a. s

    b. s^2

    c. σ

    d. σ^2

13. Consider a distribution with N = 100, X = 50, and $\Sigma(X - \overline{X})^2$ = 9875. If 10 new scores of 50 were added to the distribution, what would happen to the sum of squares?

    a. the value would increase
    b. the value would decrease
    c. the value would remain unchanged
    d. cannot be determined from information given

For questions 14 - 22 use the following sample data:

X: 10, 12, 6, 8, 9, 11, 13, 13, 5, 0, 1

14. What is the mean?

    a.   8.0
    b.   11.0
    c.   13.5
    d.   6.0

15. What is the median?

    a.   8
    b.   9
    c. 10
    d. 11

16. What is the mode?

    a.   0
    b.   6
    c. 13
    d.   9

17. What is the range?

    a. 0
    b. 13
    c. 1
    d. 3.6

18. What is the variance?

    a. 20.6
    b.   4.54
    c. 18.7
    d.   4.3

19. What is the value of $\Sigma\,(X - \overline{X})$?

    a. 4.54
    b. 206
    c. 1
    d. 0

20. What is the value of $\Sigma\,(X - \overline{X})^2$?

    a. 4.54
    b. 206
    c. 1
    d. 0

21. If the above data were a population set of scores, what would the standard deviation be?

    a. 4.33
    b. 6.92
    c. 4.54
    d. 10

22. The mean number of students in a classroom at school A is 32.5 and there are 25 classrooms. The mean number of students at school B is 29.6 in 12 classrooms. At school C the mean in each class is 15.3 for 10 classrooms. What is the mean number of students per classroom for all the schools combined?

    a. 1320.7
    b.   28.1
    c.   25.7
    d.   28.7

Answers: **1.** a **2.** b **3.** d **4.** b **5.** b **6.** d **7.** d **8.** f **9.** d **10.** c **11.** b **12.** c **13.** c **14.** a **15.** b **16.** c **17.** b **18.** a **19.** d **20.** b **21.** a **22.** b.

5 | THE NORMAL CURVE AND STANDARD SCORES

CHAPTER OUTLINE

I. **The Normal Curve**

    A. Important in behavioral sciences.

        1. Many variables of interest are approximately normally distributed.

        2. Statistical inference tests have sampling distributions which become normally distributed as sample size increases.

        3. Many statistical inference tests require sampling distributions which are normally distributed.

    B. Characteristics

        1. Symmetrical, bell-shaped curve.

        2. Equation:

$$Y = \frac{N}{\sigma\sqrt{2\pi}}\, e^{-(x-\mu)^2/2\sigma^2}$$

Shows that the curve is asymptotic to the abscissa; i.e., it approaches the X axis and gets closer and closer but never touches it.

3. Area contained under the normal curve:

    a. Area under the curve represents the percentage of scores contained within the area.

    b. 34.13% of scores between mean (μ) and $+1\sigma$; 13.59% of area contained between a score equal to $\mu + 1\sigma$ and a score of $\mu + 2\sigma$; 2.15% of area is between $\mu + 2\sigma$ and $\mu + 3\sigma$; and 0.13% falls beyond $\mu + 3\sigma$.

    c. Since the curve is symmetrical, the same percentages hold for scores below the mean.

II. Standard Scores (z Scores)

A. Converts raw scores into standard scores symbolized as z.

1. Definition. A standard score is a transformed score which designates how many standard deviation units the corresponding raw score is above or below the mean.

2. Equation:

$$z = (X - \mu)/\sigma \quad \textbf{for population data}$$

$$z = (X - \overline{X})/s \quad \textbf{for sample data}$$

B. Comparisons between different distributions.

1. Allows comparisons even when the units of the distributions are different.

2. Percentile ranks are possible.

C. Characteristics of z scores.

1. z scores have the same shape as the set of raw scores from which they were transformed.

2. $\mu_z = 0$. The mean of z scores equals zero.

3. $\sigma_z = 1.00$. The standard deviation of z scores equals 1.00.

D. Using z scores.

1. Finding the percentage or frequency (area) corresponding to any raw score.

$$z = (X - \mu)/\sigma$$

Use above formula to calculate z score. Then use table to determine the area under the normal curve for the various values of z.

2. Finding the raw score corresponding to a given percentage or frequency of scores in the distribution.

$$X = \mu + \sigma z$$

Use above formula substituting the value of z which designates the area under the curve one wishes, and solve for X, the raw score.

CONCEPT REVIEW

The (1) _____ curve is extremely important in the behavioral

sciences. Many (2) _____ measured have distributions which

closely approximate the normal curve. Many (3) _____ tests have

(4) _____ distributions which become (5) _____ distributed as

(6) _____ size (7) _____. Also, many inference tests require

sampling distributions which are normally distributed.

    The normal curve has certain characteristics. It is a (8) _____,

(9) _____ - (10) _____ curve. As the curve approaches the

(11) _____ axis, it is changing the Y value very (12) _____.

The curve (13) _____ the X axis and gets closer to it, but never

touches it. The curve is said to be (14) _____ to the X axis.

    In a normal curve, there is a special relationship between

(1) normal
(2) variables
(3) inference
(4) sampling
(5) normally
(6) sample
(7) increases
(8) symmetrical
(9) bell
(10) shaped
(11) X or horizontal
(12) slowly
(13) approaches
(14) asymptotic

the (15) _____ and (16) _____ with regard to the

(17) _____ contained under the curve. When a set of scores

is (18) _____ distributed, (19) _____ % of the area under the

curve is contained between the (20) _____ and a score which is

equal to (21) _____ + (22) _____; (23) _____ % of the

area is contained between a score equal to (24) _____ +

(25) _____ and a score of (26) _____ + (27) _____;

(28) _____ % of the area is contained between (29) _____

+ (30) _____ and (31) _____ + (32) _____; and

(33) _____ % of the area exists beyond (34) _____ +

(35) _____.

    Since the curve is symmetrical, the same percentages hold for

scores (36) _____ the mean. In graphing a normal curve

(37) _____ is plotted on the vertical axis, and the percentages

above represent the (38) _____ of (39) _____ contained

within the area.

    One can use this information to determine the frequency of

scores above or below a given score. For example, (40) _____

% of the scores would fall below the mean of a normal distri-

bution.

Standard Scores

    The main reason for standard scores is to allow one to

(41) _____ a score against a reference group. A z standard

---

(15) mean
(16) standard
     deviation
(17) area
(18) normally
(19) 34.13
(20) mean

(21) μ
(22) 1σ
(23) 13.59
(24) μ
(25) 1σ
(26) μ
(27) 2σ
(28) 2.15
(29) μ
(30) 2σ
(31) μ
(32) 3σ
(33) 0.13
(34) μ
(35) 3σ

(36) below

(37) frequency

(38) percentage
(39) scores

(40) 50

(41) compare

score is a (42) _____ score which designates how many

(43) _____ units the corresponding (44) _____

score is above or below the (45) _____. In equation

form, z = ((46) _____ - (47) _____)/(48) _____

for population data. In conjunction with the normal curve the z

score allows one to determine the (49) _____ or (50) _____

of scores which fall above or below any score in the distribution.

Consider a population set of scores where μ = 30 and σ = 5.

A raw score value of 40 would have a value of z = ((51) _____

- (52) _____)/(53)= (54) _____. Therefore, a raw score

value of 40 would have a percentile rank of (55) _____.

An important use of z scores is to be able to (56) _____ scores

that are not otherwise directly comparable. By converting your

height and weight to (57) _____ scores, you could compare

your (58) _____ standing on height and weight.

There are three characteristics of z scores worth remem-

bering. First, z scores have the same (59) _____ as the set

of raw scores. Second, the mean of the z scores (symbolized by

(60) _____) equals (61) _____. Third, the (62) _____

of z scores equals (63) _____.

There is a (64) _____ of z scores which gives the

(65) _____ of area under the standard normal curve for any

given (66) _____ score. In the text the first column (A) of the z

table contains the (67) _____ score. Column B lists the

(42)	transformed
(43)	standard deviation
(44)	raw
(45)	mean
(46)	X
(47)	μ
(48)	σ
(49)	number
(50)	percentage
(51)	40
(52)	30
(53)	5
(54)	2.00
(55)	97.72
(56)	compare
(57)	z
(58)	relative
(59)	shape
(60)	μ_z
(61)	0
(62)	standard deviation
(63)	1.00
(64)	table (Table A)
(65)	proportion
(66)	z
(67)	z

(68) _____ of the total area between a given z score and the   (68) proportion

(69) _____. Column C lists the proportion of the total area   (69) mean

which exists (70) _____ the z score. One can use the table to   (70) beyond

find the (71) _____ rank of a given (72) _____ score.   (71) percentile
                                                                 (72) raw

Assume that a college entrance exam was normally distributed

with μ = 400 and σ = 100. To determine the percentile rank

of a score of 450 you must first calculate the (73) _____   (73) z

(74) _____. To do this subtract (75) _____ from   (74) score
                                                           (75) 400

(76) _____ and divide the result by (77) _____. The result   (76) 450
                                                                       (77) 100

is z = (78) _____. The next step is to find the (79) _____   (78) 0.50
                                                                       (79) proportion

of scores less than of a z score of (80) _____. Going to Col-   (80) 0.50

umn B of the table, we find the proportion of area between the

(81) _____ and a value of 0.50 equals (82) _____. To this   (81) mean
                                                                      (82) .1915

value we must add (83) _____ to take into account the area   (83) .5000

(84) _____ the mean. Thus, the proportion of scores below a   (84) below

college entrance exam score of 450 is (85) _____ +   (85) .5000

(86) _____ = .6915. To convert this proportion to a percentile   (86) .1915

rank, (87) _____ the proportion by (88) _____. The percen-   (87) multiply
                                                                       (88) 100

tile rank of 450 is (89) _____. Another way to solve the same   (89) 69.15

problem is to go to Column (90) _____ which is the area   (90) C

(91) _____ the z score and subtract that value from   (91) beyond

(92) _____ and then multiply by (93) _____. This method   (92) 1.00
                                                                     (93) 100

works only when the raw score is greater than the (94) _____.   (94) mean

If one had a z score below the mean it would have a

(95) _____ value. The table only gives (96) _____ values.

Since the normal curve is symmetrical, to find the percentile rank

of a negative z score, one finds the (97) _____ beyond the

corresponding (98) _____ z score and multiplies by

(99) _____.

(95)	negative
(96)	positive
(97)	area
(98)	positive
(99)	100

Sometimes one wishes to find the raw score at a given

percentile from a normal distribution. Considering again the

college entrance test, ($\mu = 400$; $\sigma = 100$), how does one

determine the raw score at the 75th percentile? In the z equation

the variable (100) _____ stands for the raw score. We must

do some algebra and solve for X. The equation for the raw score

of a percentile rank is X = (101) _____ + (102) _____ x

(103) _____. An easy way to solve such a problem is to

recognize that at the 75th percentile the proportion of the area

beyond this point is (104) _____. Going to Column C of the

table, we locate the area closest to .2500 and note this point has

a z score equal to (105) _____ in Column A. Solving the

equation we have X = (106) _____ + (107) _____ x

(108) _____, which equals (109) _____.

(100)	X
(101)	μ
(102)	σ
(103)	z
(104)	.2500
(105)	0.67
(106)	400
(107)	100
(108)	.67
(109)	467

EXERCISES

For problems 1 through 8 use the following information:

In a population survey of patients in a rehabilitation hospital, the mean length of stay in the hospital was 12.0 weeks with a standard deviation equal to 1.0 week. The distribution was normally distributed.

1. Out of 100 patients how many would you expect to stay longer than 13 weeks?

2. What is the percentile rank of a stay of 11.3 weeks?

3. What percentage of patients would you expect to stay between 11.5 weeks and 13.0 weeks?

4. What percentage of patients would you expect to be in longer than 12.0 weeks?

5. How many times out of 10,000 would you expect a patient selected at random to remain in the hospital longer than 14.6 weeks?

6. What proportion of patients are likely to be in less than 9.7 weeks?

7. What is the length of stay at the 90th percentile?

8. What is the length of stay at the 50th percentile?

9. On one college aptitude test with a mean of $\mu = 100$ and a standard deviation of $\sigma = 16$, a student achieved a score of 124. The same student took a different test which had a mean of $\mu = 50$ and a standard deviation of $\sigma = 10$. On the second test the student achieved a score of 65. On which test did the student do better?

10. If the mean height of college males is 70 inches with a standard deviation of 3 inches, what percentage of college males would be between 6' and 6'4"? Assume a normal distribution.

11. Using the information in problem 10, what height would someone have to be in order to be in the 99th percentile?

12. Using the data in problem 10, what is the height below which the shortest 2.5% of the college males fall and what is the height above which the tallest 2.5% fall?

13. A surgeon is experimenting with a new technique for implanting artificial blood vessels. Using this technique with a great many operations, the mean time before clotting of an artificial blood vessel has been 32.5 days with a standard deviation of 2.6 days. The following data were obtained on four operations.

Operation	Time to Clot
A	40
B	31
C	29
D	36

a. What are the z scores for the four operations?
b. What is the percentile rank for each of the four operations? Assume a normal distribution.
c. How long would a vessel have to stay open to be in the 95th percentile? Assume a normal distribution.

14. Given the following z scores, find the area below z:

a. 1.68
b. -0.45
c. -1.96
d. -.52
e. 2.58

15. Assuming that you wished to have the highest possible score on an exam relative to the other scores; would you rather have a score of 70 on a test with a mean of 60 and a standard deviation of 5.2 or a score of 81 on a test with a mean of 70 and a standard deviation of 7.1?

16. What is the percentile rank for each of the following z scores? Assume a normal distribution.

a. 1.23
b. 0.89
c. -0.46
d. -1.00

17. What z scores correspond to the following percentile ranks? Assume the scores are normally distributed.

a. 50
b. 46
c. 96
d. 75
e. 34
f. 4

Answers: 1. 16 to nearest whole number   **2.** 24.20%   **3.** 53.28%   **4.** 50%   **5.** 47
6. .0107   **7.** 13.28 weeks   **8.** 12 weeks   **9.** the student did equally well on both tests

10. 22.86%   **11.** 76.99 inches   **12.** shortest, less than 64.12 inches; tallest, greater than 75.88 inches   **13a.** A = 2.88; B = -.58; C = -1.35; D = 1.35   **13b.** A = 99.80%; B = 28.10%; C = 8.85%; D = 91.15%   **13c.** 36.78 days (using z = 1.645)   **14a.** .9535   **14b.** .3264   **14c.** .0250   **14d.** .3015   **14e.** .9951   **15.** 81 has a z score of 1.55, while the 68 has a z score of 1.54. The 70 is the higher score relative to the other.   **16a.** 89.07   **16b.** 81.33   **16c.** 32.28   **16d.** 15.87   **17a.** 0.00   **17b.** -0.10   **17c.** 1.75   **17d.** 0.67   **17e.** -0.41   **17f.** -1.75.

TRUE-FALSE QUESTIONS

T F   1. The normal curve is a symmetric, bell-shaped curve.

T F   2. The area under the normal curve represents the percentage of scores contained within the area.

T F   3. It is impossible to have a z score of 23.5.

T F   4. A z transformation will allow comparisons to be made when units of distributions are different.

T F   5. If the original raw score distribution is not normally distributed, the mean of the z transformation scores of the raw data will not equal 0.

T F   6. In the standard normal curve, 13.59% of the scores will always be contained between the mean (μ) and $+1\sigma$.

T F   7. In a plot of the normal curve, frequency is plotted on the X axis.

T F   8. The standard deviation of the z distribution is always equal to 1.0.

T F   9. The area beyond a z score of +2.58 is .005.

T F 10. One cannot reasonably do z transformations on ratio data.

T F 11. The normal curve never touches the X axis.

T F 12. To do a z transformation, one must know only the population mean and the value of the raw score to be transformed.

T F 13. To calculate the score at the 97.5th percentile, one would apply the formula $X = \mu + (\sigma)(1.96)$.

T F 14. The z score and the z distribution are the same thing.

SELF-QUIZ

1. If a distribution of raw scores is negatively skewed, transforming the raw scores into z scores will result in a _____ distribution.

    a. normal
    b. bell-shaped
    c. positively skewed
    d. negatively skewed

2. The mean of the z distribution equals _____.

    a. 0
    b. 1
    c. N
    d. depends on the raw scores

3. The standard deviation of the z distribution equals _____.

    a. 0
    b. 1
    c. the variance of the z distribution
    d. b and c

4. $\sum(z - \mu_z)$ equals _____.

    a. 0
    b. 1
    c. the variance
    d. cannot be determined

5. The proportion of scores less than z = 0.00 is _____.

    a. 0
    b. 0.50
    c. 1.00
    d. -0.50

6. In a normal distribution the z score for the mean equals _____.

    a. 0
    b. the z score for the median
    c. the z score for the mode
    d. all of the above

7. In a normal distribution approximately _____ of the scores will fall within 1 standard deviation of the mean.

    a. 14%
    b. 95%
    c. 70%
    d. 83%

8. Would you rather have an income (assume a normal distribution and you are greedy) _____.

    a. with a z score of 1.96
    b. in the 95th percentile
    c. with a z score of -2.00
    d. with a z score of 0.000

9. How much would your income be if its z score value was 2.58?

    a. $10,000
    b. $ 9,999
    c. $ 5,000
    d. cannot be determined from information given

10. Which of the following z scores represent(s) the most extreme value in a distribution of scores assuming they are normally distributed?

    a. 1.96
    b. .0001
    c. -.0002
    d. -3.12

11. What is the percentile rank of a z score of -0.47?

    a. 31.92
    b. 18.08
    c. 50.00
    d. 47.00
    e. 0.06

12. A standardized test has a mean of 88 and a standard deviation of 12. What is the score at the 90th percentile? Assume a normal distribution.

    a. 90.00
    b. 112.00
    c. 103.36
    d. 91.00

13. On a test with a population mean of 75 and standard deviation equal to 16, if the scores are normally distributed, what is the percentile rank of a score of 56?

    a. 58.30
    b.   0.00
    c. 25.27
    d. 38.30
    e. 11.70

14. On the test referred to in problem 13, what percentage of scores fall below a score of 83.8?

    a. 55.00
    b. 79.12
    c. 20.88
    d. 29.12
    e. 70.88

15. Using the same data as in problem 13, what percentage of scores fall between 70 and 80?

    a. 75.66
    b. 70 23
    c. 24.34
    d. 23.57
    e. 12.17

Answers: 1. d   **2.** a   **3.** d   **4.** a   **5.** b   **6.** d   **7.** c   **8.** a   **9.** d   **10.** d
11. a   **12.** c   **13.** e   **14.** e   **15.** c.

6 | LINEAR REGRESSION

CHAPTER OUTLINE

I. Introduction.

A. <u>Linear regression.</u> This topic deals with predicting scores of one distribution using information known about scores on a second distribution. For example, one might predict your height if they knew your weight and the nature of your relationship between height and weight from a sample of other people.

B. <u>Correlation.</u> This refers to the magnitude and direction of the relationship between two variables.

II. Linear Relationships.

A. <u>Linear.</u> A linear relationship between two variables is one in which the relationship between two variables can accurately be represented by a straight line.

B. <u>Curvilinear.</u> When a curved line fits a set of points better than a straight line it is called a curvilinear association or relationship.

C. <u>Scatter plots.</u> A scatter plot is a graph of paired X (one variable score) and Y (another variable score) values. By visually examining the graph one can get a good idea of the nature of the relationship between the two variables (i.e., linear or not).

III. Straight Line Equation.

A. <u>General equation.</u> $Y = bX + a$ where a = the Y intercept and b = the slope of the line.

B. <u>Slope of the straight line equation (b).</u> The slope tells us how much the Y score changes for each unit change in the X score. In equation form:

$$b = \text{slope} = (Y_2 - Y_1)/(X_2 - X_1)$$

The slope is a constant value.

C. <u>Y intercept (a).</u> The Y intercept is the value of Y where the line intersects the Y axis. It is the value of Y when $X = 0$.

D. Relationships.

   1. Positive relationships. This indicates that there is a direct relationship between the variables. Higher values of X are associated with higher values of Y and vice versa.

   2. Negative relationships. This exists when there is an inverse relationship between X and Y. Low values of X are associated with high values of Y and vice versa.

   3. Perfect relationship. This occurs when all the pairs of points fall on a straight line.

   4. Imperfect relationships. This is when a positive or negative, relationship exists but all of the points do not fall on the line.

IV. Least-Squares Regression Line for Prediction.

A. <u>Least-squares criterion.</u> In an imperfect relationship no single straight line will hit all the points. We pick the line that will minimize the total errors of prediction, i.e., construct the one line that minimizes $\sum (Y - Y')^2$ where Y' is the predicted value of Y for any value of X.

B. Constructing the regression line of Y on X:

$$Y' = b_Y X + a_Y$$

where

$$b_Y = \frac{\Sigma\ XY - \dfrac{(\Sigma\ X)(\Sigma\ Y)}{N}}{\Sigma\ X^2 - \dfrac{(\Sigma\ X)^2}{N}}$$

$$a_Y = \overline{Y} - b_Y X$$

C. <u>Use of regression equation.</u> For a given value of X, simply plug that value in the equation and solve for Y' using the regression constants b_Y and a_Y.

D. <u>Regression of X on Y.</u> In some cases one wants to predict X given a value of Y. The equation of interest becomes:

$$X' = b_X Y + a_X$$

where

$$b_X = \frac{\Sigma\ XY - \dfrac{(\Sigma\ X)(\Sigma\ Y)}{N}}{\Sigma\ Y^2 - \dfrac{(\Sigma\ Y)^2}{N}}$$

$$a_X = \overline{X} - b_X Y$$

V. Prediction Errors. When relationships between X and Y variables are imperfect, there will be prediction errors.

A. <u>Standard error of estimate.</u> ($s_{Y|X}$). Quantifying the magnitude of the error involves computing the standard error of estimatc symbolized $s_{Y|X}$. The standard error is much like the standard deviation.

1. Definition. Gives a measure of the average deviation of the prediction errors about the regression line.

2. Equation for standard error of estimate.

$$s_{Y|X} = \sqrt{\frac{SS_Y - \dfrac{[\Sigma\ XY - (\Sigma\ X)(\Sigma\ Y)/N]^2}{SS_X}}{N-2}}$$

3. Interpretation. The larger the value of $s_{Y|X}$, the less confidence one has in the prediction of Y given X. The smaller the value of $s_{Y|X}$, the more likely the prediction will be accurate. If one constructed two parallel lines to the regression line at distances of $\pm 1 s_{Y|X}$, $\pm 2 s_{Y|X}$, and $\pm 3 s_{Y|X}$, one would find about 68%, 95%, and 99% of the scores would fall between the lines respectively.

B. <u>Other errors.</u> One must be careful of sources of errors in making predictions. There are two major considerations in making predictions.

1. Linearity. The original relationship needs to be linear for accurate prediction using linear regression.

2. Prediction in the range. Generally one uses a sample to generate the data for calculating the regression constants (b_Y and a_Y). Predictions of Y should be based on values of' X within the range of the sample upon which the constants are based.

CONCEPT REVIEW

It is often useful to use knowledge of one variable to predict

a likely value on a second variable. If there is a (1) _____   (1) relationship

between two variables, we can use knowledge of this

relationship for (2) _____. The name of this topic which covers   (2) prediction

this material for linear relationships is (3) _____ (4) _____.   (3) linear
                                                                            (4) regression

The easiest way to determine if a relationship exists between two

variables is to (5) _____ the variables on a (6) _____.   (5) plot
                                                                     (6) graph
Such a plot is called a (7) _____ (8) _____. A scatter plot   (7) scatter

is a graph of (9) _____ X and Y scores. When a (10) _____ line accurately describes the relationship between two variables, the relationship is called (11) _____. Not all relationships are linear. Those that are not are called (12) _____. In these cases a (13) _____ line fits the (14) _____ better than a (15) _____ line. Although a (16) _____ solution is sometimes used for prediction, it is more common to predict (17) _____ from the (18) _____ of the straight line. The general form of the equation for a straight line is:

Y = (19) _____ times (20) _____ + (21) _____,

where (22) _____ = the Y intercept and (23) _____ = the slope of the line.

The (24) _____ is the value of Y where the line intersects the (25) _____. Thus, it is the value of (26) _____ when X = (27) _____

The slope of a line measures its (28) _____ of (29) _____. The slope tells how much the (30) _____ score changes for each (31) _____ change in the (32) _____ score. In straight line functions, the slope has a (33) _____ value for any points on the line. In conceptual terms the equation for the slope is:

(8) plot
(9) paired
(10) straight

(11) linear

(12) curvilinear
(13) curved
(14) points
(15) straight
(16) graphic

(17) Y
(18) equation

(19) b
(20) X
(21) a

(22) a
(23) b

(24) Y intercept

(25) Y axis or ordinate
(26) Y
(27) 0
(28) rate
(29) change
(30) Y

(31) unit
(32) X
(33) constant

$$b = \frac{\Delta \ (34) \ _____}{\Delta \ (35) \ \_\_\_\_\_} = \frac{(36) \ \_\_\_\_\_ - (37) \ \_\_\_\_\_}{(38) \ \_\_\_\_\_ - (39) \ \_\_\_\_\_}$$

(34) Y
(35) X
(36) Y_2
(37) Y_1
(38) X_2
(39) X_1

If one had two pairs of points (10, 20) and (15, 30), the slope for

the line connecting these points would be:

$$b = \frac{(40) \ \_\_\_\_\_ - (41) \ \_\_\_\_\_}{(42) \ \_\_\_\_\_ - (43) \ \_\_\_\_\_} = (44) \ \_\_\_\_\_$$

(40) 30
(41) 20
(42) 15
(43) 10
(44)   2

Relationships between two variables may be either (45) _____

or (46) _____. If the relationship is positive, the slope is

(47) _____. If the slope is (48) _____ the relationship is

(49) _____. In a positive relationship higher values of

(50) _____ are associated with (51) _____ values of

(52) _____. In a negative relationship (53) _____ values of

(54) _____ are associated with (55) _____ values of

(56) _____. On a graph, a negative slope would run down-

ward from (57) _____ to (58) _____. In a negative relation-

ship as X (59) _____, Y (60) _____. In a positive relation-

ship, as X (61) _____ Y (62) _____.

In an (63) _____ relationship, all of the points do not fall on

the regression line. In an imperfect relationship one constructs

the line which (64) _____ errors of (65) _____ according to

a (66) _____ - (67) _____ criterion. This is called the

(45) positive
(46) negative

(47) positive
(48) negative
(49) negative

(50) X
(51) higher
(52) Y
(53) lower
(54) X
(55) higher
(56) Y

(57) left
(58) right
(59) increases
(60) decreases
(61) increases
(62) increases
(63) imperfect

(64) minimizes
(65) prediction
(66) least

(68) _____ -(69) _____ (70) _____ (71) _____.

The (72) _____ (73) _____ between the regression line and

each (74) _____ represents the (75) _____ in

prediction. (76) _____ equals the predicted Y value

and (77) _____ equals the actual value of Y. (78) _____ -

(79) _____ equals the error for each point. The least-

squares regression line minimizes (80) _____.

(67)	squares
(68)	least
(69)	squares
(70)	regression
(71)	line
(72)	vertical
(73)	distance
(74)	point
(75)	error
(76)	Y'
(77)	Y
(78)	Y
(79)	Y'
(80)	$\Sigma (Y - Y')^2$

Constructing the Regression Line

The terms (81) _____ and (82) _____ are called regress-

ion (83) _____. The regression line for predicting Y given X is

constructed by computing values for (84) _____ and

(85) _____. The computational formula for computing b_Y is:

(81)	b_Y
(82)	a_Y
(83)	constants
(84)	b_Y
(85)	a_Y

$$b_Y = \frac{(86)\_\_\_ - \frac{(87)\_\_\_\ (88)\_\_\_}{(89)\_\_\_}}{(90)\_\_\_ - \frac{(91)\_\_\_}{(92)\_\_\_}}$$

(86)	ΣXY
(87)	(ΣX)
(88)	(ΣY)
(89)	N
(90)	ΣX^2
(91)	$(\Sigma X)^2$
(92)	N

N is equal to the number of (93) _____ (94) _____.

(93)	paired
(94)	scores

The a_Y regression constant is given by the equation:

$$a_Y = (95)\ \_\_\_ - (96)\ \_\_\_\ (97)\ \_\_\_$$

(95)	$\overline{Y}$
(96)	b_Y
(97)	$\overline{X}$

Since we need to know the value of b_Y to determine the a_Y

constant, we first find (98) _____, then (99) _____. Once

they are both found they are substituted into the (100) _____

(98)	b_Y
(99)	a_Y
(100)	regression

equation. The above regression constants are for the values of the

regression line of (101) _____ on (102) _____. It is some-

(101) Y
(102) X

times of interest to predict X given Y. This is called the regression

line of (103) _____ on (104) _____. The linear regression

(103) X
(104) Y

equation for predicting X given Y is:

(105) _____ = (106) _____ times (107) _____ + a_X

(105) X'
(106) b_X
(107) Y

This regression line is constructed by calculating values for

(108) _____ and (109) _____. The computational formula

(108) b_X
(109) a_X

for b_X is:

$$b_X = \frac{(110)\ \_\_\_\_ - \dfrac{(111)\ \_\_\_\_\ (112)\ \_\_\_\_}{(113)\ \_\_\_\_}}{(114)\ \_\_\_\_ - \dfrac{(115)\ \_\_\_\_}{(116)\ \_\_\_\_}}$$

(110) $\Sigma\,XY$
(111) $(\Sigma\,X)$
(112) $(\Sigma\,Y)$
(113) N
(114) $\Sigma\,Y^2$
(115) $(\Sigma\,Y)^2$
(116) N

The equation for a_X is:

a_X = (117) _____ - (118) _____ (119) _____

(117) $\overline{X}$
(118) $\underline{b_X}$
(119) $\overline{Y}$

The regression line of Y on X will equal the regression line of X

on Y only when the relationship is (120) _____.

(120) perfect

Measuring Prediction Errors

Quantifying prediction errors involves computing the

(121) _____ (122) _____ of (123) _____ which is sym-

(121) standard
(122) error
(123) estimate

bolized by (124) _____. The standard error of estimate gives

a measure of the (125) _____ deviation of the (126) _____

(124) $s_{Y|X}$
(125) average

errors about the (127) _____ (128) _____. The conceptual

(126) prediction
(127) regression
(128) line

formula for $s_{Y|X}$ is:

$$s_{Y|X} = \sqrt{(129)\ \underline{\hspace{1cm}}\ /\ (130)\ \underline{\hspace{1cm}}}$$

(129) $\sum (Y - Y')^2$
(130) $N - 2$

The computational formula is:

$$s_{Y|X} = \sqrt{\dfrac{(131)\ \underline{\hspace{1cm}} - \dfrac{(132)\ \underline{\hspace{1cm}}}{(133)\ \underline{\hspace{1cm}}}}{(134)\ \underline{\hspace{1cm}} - (135)\ \underline{\hspace{1cm}}}}$$

(131) SS_Y
(132) $[\sum XY - (\sum X)(\sum Y)/N]^2$
(133) SS_X
(134) N
(135) 2

for predicting (136) _____ given (137) _____. The

(136) Y
(137) X
(138) Y

standard error of estimate is computed over all (138) _____

scores. For it to be meaningful one assumes that the

(139) _____ of Y remains (140) _____ as one goes from

(139) variability
(140) constant
(141) X

one (141) _____ score to the next. This assumption is called

the assumption of (142) _____. In general one would expect

(142) homoscedas-
ticity

to find (143) _____ of the points to fall within $\pm 1 s_{Y|X}$ of the

(143) 68%

regression line, 95% of the points to fall within $\pm$(144) _____

(144) 2

$s_{Y|X}$, and (145) _____ % to fall within $\pm$(146) _____.

(145) 99
(146) $3\,s_{Y|X}$

In general it is appropriate to use linear regression to predict

values only if the (147) _____ is (148) _____. It is also

(147) relationship
(148) linear
(149) computation
or sample

important that the basic (149) _____ group be representative of

the (150) _____ group. In other words, the data collected to

(150) prediction

compute the (151) _____ (152) _____ should be a

(151) regression
(152) constants

(153) _____ sample from the (154) _____ of interest.

(153) random
(154) population

Finally, the linear regression equation is properly used just for the

(155) _____ of the variable upon which it is based. This is   (155) range

because we do not know if data outside the range of our sample

continues to be a (156) _____ (157) _____.   (156) linear
          (157) relationship

EXERCISES

1. X represents aptitude test scores and Y represents grade point average in college. If the least-squares regression line for the relationship between these two variables is $Y' = .005X + 1.2$, what GPA would you predict for people who scored each of the following scores on the aptitude test?

    a. 159
    b. 300
    c. 500
    d. 550

2. Draw a graph of aptitude test score versus grade point average and construct the regression line for the line $Y' = .005X + 1.2$.

3. A professor wanted to predict final exam scores from midterm exam scores. He used data from several different professors teaching the same class. He obtained the following data:

Midterm Scores:	83,	62,	72,	85,	85
Final Exam Scores:	89,	58,	70,	92,	84

    What are the values for each of the following?

    a. $\sum X$.

    b. $\sum X^2$.

    c. N.

    d. $\sum Y$.

    e. $\sum Y^2$.

f. $(\sum X)^2$.

g. $(\sum Y)^2$.

h. $\sum XY$.

i. b_Y.

j. a_Y.

k. If the professor's class score on the midterm was 77.4, what score would you predict the class would receive on the final exam?

l. What is the value of $s_{Y|X}$?

4. A hospital administrator wanted to predict the number of patients her hospital would admit in 1990. The following data were obtained from past records:

Year:	1960,	1965,	1970,	1975,	1980
Number of Admissions:	812,	983,	1127,	904,	1768

a. What would the best prediction be for the number of admissions expected in 1990?
b. What serious caution should the administrator be aware of when making her prediction?

5. A psychologist wanted to use a locus of control test to predict scores on a depression scale. The following data were summarized for the relationship between the locus of control and depression scale:

$$\sum X = 62, \quad \sum X^2 = 1022, \quad \sum Y = 70, \quad \sum Y^2 = 1234, \quad \sum XY = 1107, \quad N = 4$$

a. What is the value of b_Y?
b. What is the value of a_Y?
c. What would the psychologist predict for the score on the depression scale if a client scored an 18 on the locus of control scale?

6. Consider the following set of data points:

X	2	4	8	14	20	23	25
Y	2	6	14	20	12	9	7

a. Construct a scatter plot of the points.
b. Is it appropriate to use a least-squares 1inear regression line to predict Y from X in this case? Why or why not?

7. Consider the following set of points for variable X and variable Y:

X	21	29	33	40	50
Y	34	36	42	45	58

Assume the relationship is linear in answering the following questions.

a. What is the value of the regression constant b_X for predicting X given Y?
b. What is the value of the regression constant a_X for predicting X given Y?
c. What value of X would you predict for a value of Y = 37?
d. What value of X would you predict for a value of Y = 55?

8. What is the linear regression equation for predicting Y given X for the following pairs of scores:

X	9	15	25	27	42	50	30
Y	14	11	5	5	0	-8	1

Answers: 1a. 2.00   **1b.** 2.70   **1c.** 3.70   **1d.** 3.95
2.

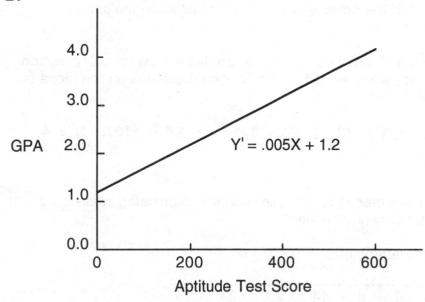

$Y' = .005X + 1.2$

3a. 387   **3b.** 30,367   **3c.** 5   **3d.** 393   **3e.** 31,705   **3f.** 149,769   **3g.** 154,449
3h. 30,983   **3i.** 1.367   **3j.** -27.198   **3k.** 78.6   **3l.** 3.285   **4a.** 1852   **4b.** When predicting beyond the range of the sample data, one cannot be sure the nature of the relationship will remain the same.   **5a.** 0.3607   **5b.** 11.91   **5c.** 18.4

6a.

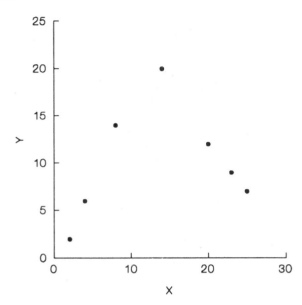

6b. no, the relationship is curvilinear **7a.** $b_X = 1.125$ **7b.** $a_X = -13.775$ **7c.** $X' = 27.85$ **7d.** $X' = 48.10$ **8.** $Y' = -0.499X + 18.126.$

TRUE-FALSE QUESTIONS

T F 1. The easiest way to determine if a relationship is linear is to calculate the regression line.

T F 2. In a linear relationship all the points must fall on a straight line.

T F 3. In a perfect linear relationship all the points must fall on a straight line.

T F 4. The slope of a line is a measure of its rate of change.

T F 5. In a straight line the slope approaches zero as the line comes near the point X, Y.

T F 6. In an inverse relationship as one variable gets larger the other variable gets smaller.

T F 7. In regression analysis we are only concerned with perfect as opposed to imperfect relationships.

T F 8. If we minimize $\sum (Y-Y')^2$, we will minimize the total error of prediction.

T F   9. The value a_Y is the X axis intercept for minimizing errors in Y.

T F 10. Generally, one can use the same regression equation for predicting Y given X as for X given Y.

T F 11. If the relationship between two variables is perfect the standard error of estimate equals 0.

T F 12. If the standard error of estimate for relationship 1 equals 5.26 and for relationship 2 it equals 8.01 then we can reasonably infer that relationship 2 is less perfect than relationship 1.

T F 13. It is impossible to have a negative value for the standard error of estimate.

T F 14. In general one is less confident in predictions of Y when the value of X used for the prediction is outside the range of the original data used to construct the regression line.

T F 15. If the regression line is parallel to the X axis then the slope of the regression line equals 0.

T F 16. The regression line will always go through the point $\overline{X}, \overline{Y}$. (Try it and see.)

Answers: 1. F   **2.** F   **3.** T   **4.** T   **5.** F   **6.** T   **7.** F   **8.** T   **9.** F   **10.** F
11. T   **12.** T   **13.** T   **14.** T   **15.** T   **16.** T.

SELF-QUIZ

1. If $s_{Y|X} = 0.0$ the relationship between the variables is _____.

    a. perfect
    b. imperfect
    c. curvilinear
    d. unknown

2. $\Sigma (Y - Y')$ equals _____.

    a. 0
    b. 1
    c. cannot be determined from information given
    d. who cares

3. $\sum (Y - Y')^2$ represents _____.

   a. the standard deviation
   b. the variance
   c. the standard error of estimate
   d. the total error of prediction

4. In a particular relationship N = 80. How many points would you expect on the average to find within $\pm 1 s_{Y|X}$ of the regression line?

   a. 40
   b. 80
   c. 54
   d. 0

5. What would you predict for the value of Y for the point where the value of X is $\overline{X}$?

   a. cannot be determined from information given
   b. 0
   c. 1
   d. $\overline{Y}$

6. If the value of $s_{Y|X} = 4.00$ for relationship A and $s_{Y|X} = 5.25$ for relationship B, in which relationship would you have the most confidence in a particular prediction?

   a. A
   b. B
   c. it makes no difference
   d. cannot be determined from information given

7. If b_Y is negative, higher values of X are associated with _____.

   a. lower values of X'
   b. higher values of Y
   c. higher values of (Y - Y')
   d. lower values of Y

8. Which of the following statement(s) is (are) an important consideration(s) in applying linear regression techniques?

   a. the relationship should be linear
   b. both variables must be measured in the same units
   c. predictions for Y should be within the range of the X variable in the sample
   d. a and c

9. In the regression equation $Y' = X$, the Y-intercept is _____.

    a. $\overline{X}$
    b. $\overline{Y}$
    c. 0
    d. 1

10. If the value for a_Y is negative, the relationship between X and Y is _____.

    a. positive
    b. negative
    c. inverse
    d. cannot be determined from information given

11. If $b_Y = 0$, the regression line is _____.

    a. horizontal
    b. vertical
    c. undefined
    d. at a 45° angle to the X axis

12. The least-squares regression line minimizes _____.

    a. s
    b. $s_{Y|X}$
    c. $\sum (Y - \overline{Y})^2$
    d. $\sum (Y - Y')^2$
    e. b and d

13. The points (0,5) and (5,10) fall on the regression line for a perfect positive linear relationship. What is the regression equation for this relationship?

    a. $Y' = X + 5$
    b. $Y' = 5X$
    c. $Y' = 5X + 10$
    d. cannot be determined from information given

14. For the following points what would you predict to be the value of Y' when X = 19? Assume a linear relationship.

X	6	12	30	40
Y	10	14	20	27

    a. 16.35
    b. 24.69
    c. 22.00
    d. 17.75

15. If $N = 8$, $\sum X = 160$, $\sum X^2 = 4656$, $\sum Y = 79$, $\sum Y^2 = 1309$, and $\sum XY = 2430$, what is the value of b_Y?

    a.   .9217
    b. -1.8010
    c.   .5838
    d.   .7922

16. What is the slope for the points $X_1 = 30$, $Y_1 = 50$ and $X_2 = 25$ and $Y_2 = 40$?

    a.   2.00
    b.   .50
    c. -2.00
    d.   -.50

17. If the regression equation for a set of data is $Y' = 2.650X + 11.250$ then the value of Y' for X = 33 is _____.

    a.   87.45
    b. 371.25
    c.   98.70
    d.   76.20

18. If $\bar{X} = 57.2$, $\bar{Y} = 84.6$, and $b_Y = .37$, the value of $a_Y =$ _____.

    a. 141.80
    b. -25.90
    c.   63.44
    d.   27.40

19. If the regression line for predicting X given Y were X' = 103Y + 26.2, what would the value of X' be if Y = 0.2?

   a.   129.2
   b.   25.8
   c.   5.2
   d.   46.8

Answers: 1. a   **2.** a   **3.** d   **4.** c   **5.** d   **6.** a   **7.** d   **8.** d   **9.** c   **10.** d
11. a   **12.** e   **13.** a   **14.** a   **15.** c   **16.** a   **17.** c   **18.** c   **19.** d.

7

CORRELATION

CHAPTER OUTLINE

I. Correlational Concepts

A. <u>Definition.</u> Correlation is a measure of the direction and degree of relationship that exists between two variables.

B. <u>Correlation coefficient.</u> Expresses quantitatively the magnitude and direction of the correlation.

1. Range. Can range from +1 to -1.

2. Sign. The sign of the coefficient tells us whether the relationship is positive or negative.

3. Magnitude. The coefficient ranges from +1 to -1. Plus 1 is a perfect positive correlation, and minus 1 expresses a perfect negative relationship. A zero value of the correlation coefficient means there is no relationship between the two variables. Imperfect relationships vary between 0 and 1. They will be plus or minus depending on the direction of the relationship.

II. Pearson r and Standard Scores

A. <u>Regression line.</u> Pearson r is the slope of the least squares linear regression line when the scores are plotted as standard (z) scores.

B. <u>Distribution position.</u> Pearson r is a measure of the extent to which paired scores occupy the same position within their own distributions. Standard scores allow us to examine the relative positions of variables independent of the units of measure.

C. Calculating r.

    1. Computational formula from raw scores:

$$r = \frac{\sum XY - \frac{(\sum X)(\sum Y)}{N}}{\sqrt{\left[\sum X^2 - \frac{(\sum X)^2}{N}\right]\left[\sum Y^2 - \frac{(\sum Y)^2}{N}\right]}}$$

III. Additional Interpretation for Pearson r

A. <u>Variability of Y.</u> Pearson r can also be interpreted in terms of the variability of Y accounted for by X.

    1. Imperfect relationships. Where r = 0, knowledge of X does not help us predict Y. Best prediction of Y when r = 0 is $\overline{Y}$.

    2. Prediction of Y. Distance between a given score (Y_i) and the mean of Y scores (Y) is divisible into two parts.

        a. Deviation of Y_i = Error in prediction + deviation of Y_i accounted for by X.

$$(Y_i - \overline{Y}) \quad = \quad (Y_i - Y') \quad + \quad (Y' - \overline{Y})$$

        b. Total variability.

$$\sum (Y_i - Y)^2 = \sum (Y_i - Y')^2 + \sum (Y' - \overline{Y})^2$$

As the relationship between X and Y gets stronger, the prediction error gets smaller causing $\sum (Y_i - Y')^2$ to decrease, and $\sum (Y' - \overline{Y})^2$ to increase.

    3. New definition of r. Pearson r equals the square root of the proportion of the total variability of Y accounted for by X. For example, if r = .7 then .49 or 49% of the variability of Y is accounted for by X. This is called the explained variability.

IV. Regression Constants and Pearson r

A. <u>Regression coefficient:</u> $b_Y = r(s_Y/s_X)$

B. Regression coefficient: $b_X = r(s_X/s_Y)$
C. Regression constant a_Y: found in the usual way

V. Other correlation coefficients besides r

A. Eta(η). This is used for curvilinear relationships where Pearson r would underestimate the degree of relationship.

B. Biserial correlation coefficient. Used when one variable is measured on an interval scale and the other variable is dichotomous.

C. Phi (Φ) coefficient. Used when both variables are dichotomous.

D. Spearman rank order correlation coefficient (r_s), also called rho. Used when one or both variables are of ordinal scaling.

1. Computational equation:

$$r_s = 1 - \frac{6\sum D_i^2}{N^3 - N}$$

where D_i = difference between ith pair of ranks
N = number of pairs of ranks

2. Uses. When the data are not of either interval or ratio scaling but are of ordinal scaling, r_s can be used.

VI. Correlation and Causation

A. Correlation between X and Y does not prove causation.

B. Explanations for correlation between X and Y:

1. Correlation may be spurious

2. X causes Y

3. Y causes X

4. Third variable causes the correlation between X and Y

C. Role of experimentation. To establish that one variable is the cause of another, an experiment must be conducted by systematically varying only the causal variable and then measuring the effect on the other variable.

VII.   Multiple Regression and Multiple Correlation

A. Extension of simple regression (single predictor) to situations that involve two or more predictor variables.

B. Useful for increasing accuracy of prediction.

C. Equation for two predictor variables:

$$Y' = b_1X_1 + b_2X_2 + a$$

D. R^2 = multiple coefficient of determination = squared multiple correlation.

E. Equation of R^2 for two predictor variables.

$$R^2 = \frac{r_{YX_1}^{2} + r_{YX_2}^{2} - 2r_{YX_1} r_{YX_2} r_{X_1 X_2}}{1 - r_{X_1 X_2}^{2}}$$

CONCEPT REVIEW

(1) _____ and (2) _____ are closely related. Correlation is used to determine whether a (3) _____ exists between two variables and to establish its (4) _____ and (5) _____.

The degree of relationship may vary from being (6) _____ to (7) _____. When the relationship is perfect, the correlation is at its (8) _____ and we can exactly (9) _____ one variable from the other. As X changes, so does (10) _____. The points will all fall on a (11) _____ (12) _____. When a relationship

(1) Correlation
(2) regression
(3) relationship
(4) magnitude
(5) direction
(6) nonexistent
(7) perfect
(8) highest
(9) predict
(10) Y
(11) straight
(12) line

is nonexistent, the correlation is (13) _____, and knowing

the value of one variable doesn't help at all in (14) _____ the

other. (15) _____ relationships have intermediate levels

of correlation and (16) _____ is approximate. The same value

of X doesn't always lead to the (17) _____ value of Y.

But on the average, Y changes (18) _____ with X and we

can do a better job of(19) _____ Y with knowledge of X

than without it. The correlation (20) _____ expresses

quantitatively the (21) _____ and (22) _____ of the

relationship. The correlation coefficient can vary from

(23) _____ to (24) _____. The (25) _____ of the

coefficient tells us whether the relationship is positive

or negative. The (26) _____ part of the coefficient

tells us the magnitude of the correlation. The (27) _____ the

absolute value of the correlation coefficient, the stronger

the relationship. A correlation of (28) _____ or (29) _____

means the relationship is perfect, and a correlation of

(30) _____ means there is (31) _____ relationship

between the two variables. Imperfect relationships have mag-

nitudes varying between (32) _____ and (33) _____. They

will be (34) _____ or (35) _____ depending on the direction

of the relationship. In a positive relationship, as X (36) _____,

Y (37) _____. In a (38) _____ relationship, as X increases,

Y decreases. In the case where there is a zero correlation the

(13) lowest

(14) predicting

(15) Imperfect

(16) prediction

(17) same

(18) systematically

(19) predicting

(20) coefficient

(21) magnitude
(22) direction

(23) +1
(24) -1
(25) sign

(26) numerical

(27) higher

(28) +1
(29) -1

(30) 0
(31) no

(32) 0
(33) 1
(34) plus
(35) minus
(36) increases

(37) increases
(38) negative

regression line for predicting Y is (39) _____.

    Correlation and prediction are closely interrelated. One

definition of r is that r is the (40) _____ of the least-(41) _____

regression line when the scores are plotted as (42) _____

scores. Standard scores (z scores) are used in determining r to

allow measurement of the relationship between the two

variables which is (43) _____ of differences in (44) _____

and (45) _____.

    Pearson r is a measure of the extent to which paired

scores occupy the same (46) _____ within their own

distribution. When each pair of scores has the same (47) _____

value, the slope of the regression line equals (48) _____. As

the magnitude of the association between two variables de-

creases, the slope of the regression line when plotted in standard

(z) scores (49) _____. When there is no relationship both r and

the slope equal (50) _____.

    To calculate the value of r from raw scores, we can utilize the

following formula:

(39)	horizontal
(40)	slope
(41)	squares
(42)	standard (z)
(43)	independent
(44)	scaling
(45)	units
(46)	position
(47)	z
(48)	1
(49)	decreases
(50)	0

$$r = \cfrac{(51)\underline{\quad} - \cfrac{((52)\underline{\quad})((53)\underline{\quad})}{(54)\underline{\quad}}}{\sqrt{\left((55)\underline{\quad} - \cfrac{(56)\underline{\quad}}{(57)\underline{\quad}}\right)\left((58)\underline{\quad} - \cfrac{(59)\underline{\quad}}{(60)\underline{\quad}}\right)}}$$

(51)	$\Sigma\, XY$
(52)	$\Sigma\, X$
(53)	$\Sigma\, Y$
(54)	N
(55)	$\Sigma\, X^2$
(56)	$(\Sigma\, X)^2$
(57)	N
(58)	$\Sigma\, Y^2$
(59)	$(\Sigma\, Y)^2$
(60)	N

All these values are familiar to us by now and many of these values should already be available if we have calculated the regression line for the data.

In predicting Y from X the (61) _____ variability of Y can be divided into two parts; the variability of (62) _____ errors and the variability of (63) _____ accounted for by (64) _____.

The total variability is symbolized by (65) _____. The variability of prediction errors is symbolized by (66) _____. The variability of Y accounted for by X is symbolized by (67) _____. As the relationship gets stronger the variability of prediction errors gets (68) _____ and the variability of Y accounted for by X gets (69) _____. Therefore, the proportion of the total variability of the Y scores accounted for by X is another measure of the strength of the (70) _____. A conceptual formula for r is:

$$r = \sqrt{\frac{\Sigma \left((71)\_\_\_\_ - (72)\_\_\_\_ \right)^2}{\Sigma \left((73)\_\_\_\_ - (74)\_\_\_\_ \right)^2}}$$

Squaring both sides of the above equation gives us useful information about how important X is in explaining the (75) _____ of Y. Hence, r^2 = the (76) _____ of the total variability of (77) _____ accounted for by (78) _____.

Because of the close relationship between regression and correlation, the regression constant for the (79) _____ can be

(61) total
(62) prediction
(63) Y
(64) X
(65) $\Sigma (Y_i - \overline{Y})^2$
(66) $\Sigma (Y_i - \overline{Y'})^2$
(67) $\Sigma (Y' - \overline{Y})^2$
(68) smaller
(69) larger
(70) relationship
(71) Y'
(72) $\overline{Y}$
(73) Y_i
(74) $\overline{Y}$
(75) variability
(76) proportion
(77) Y
(78) X
(79) slope

determined from the values of (80) _____, (81) _____, and

(82) _____. The equation is:

$$b_Y = (83) \ \text{_____} \ \times \ \frac{(84) \ \text{\_\_\_\_\_}}{(85) \ \text{\_\_\_\_\_}}$$

for the slope of the regression line of Y upon X. For the regression

line of X upon Y, the formula is:

$$b_X = (86) \ \text{_____} \ \times \ \frac{(87) \ \text{\_\_\_\_\_}}{(88) \ \text{\_\_\_\_\_}}$$

     Besides Pearson r, there are other correlation coefficients. When

the data is curvilinear, one can use the correlation coefficient

(89) _____. Pearson r is used when data are measured

on an (90) _____ or (91) _____ scale. If one variable

is at least (92) _____ and the other (93) _____,

the biserial correlation coefficient is used. If both variables are

dichotomous, then (94) _____ is used. Perhaps the second

most often used correlation coefficient is the (95) _____ rank

order coefficient, symbolized by (96) _____. The equation for

r_S is:

$$r_S = 1 - \frac{6 \Sigma \ D_i^2}{(97) \ \text{\_\_\_\_\_} - N}$$

where

    D_i = the (98) _____ between the ith pair of ranks

(80) r
(81) s_Y
(82) s_X

(83) r
(84) s_Y
(85) s_X

(86) r
(87) s_X
(88) s_Y

(89) eta

(90) interval
(91) ratio
(92) interval
(93) dichotomous

(94) phi

(95) Spearman

(96) r_S

(97) N^3

(98) difference

N = the number of (99) _____ of ranks.　(99) pairs

In doing correlational studies there are certain things to keep in mind. If there is not sufficient (100) _____ of both variables,　(100) range the correlation may be artifically low. Also, it is tempting to infer that correlation means causation. This inference would be unjustified without further (101) _____. There are four　(101) experiment-ation possible explanations for correlation. First, the correlation between X and Y may be (102) _____. Second, X may　(102) spurious cause Y or, third, Y may cause (103) _____. Finally, a　(103) X (104) _____ variable may cause the correlation between X　(104) third and Y.

Simple regression refers to using _____ predictor　(105) one variable. Multiple regression refers to using _____　(106) two or more predictor variables. Using _____ predictor variables　(107) two or more often increases the accuracy of predictiion. Of course, this depends on how much _____ variation the additional predictor　(108 unexplained variables can account for. _____ tells us how much variation is　(109) R^2 accounted for by all of the predictor variiables.

EXERCISES

1. What is the correlation between the following pairs of values for height and weight? Use the table to help you set up the problem.

Person	Height X	X^2	Weight Y	Y^2	XY
1	73		185		
2	68		150		
3	61		125		
4	65		130		
5	70		162		
6	69		170		

2. The phone rings and the President is calling you. He says, "I hear you are studying correlation. Can you please tell me how much of the variability of my popularity rating is explained by the inflation rate?" He kindly supplies you with the following data:

Inflation rate	10	12	8	6	14
Popularity rating	70	75	85	84	48

3a. An ornithologist wants to know how much of a relationship there is between the weight gain of the female flicker during the spring and the number of eggs laid. What is the Pearson r for the data the ornithologist gathered?

Bird	1	2	3	4	5	6
Weight Gain (oz.)	2	6	6	3	10	7
Number of baby Flickers	0	1	3	3	6	5

3b. How much of the variability in the number of eggs is accounted for by knowing the amount of weight gained?

4. A teacher was interested in knowing if leadership ability was correlated with attractiveness in third graders. The teacher ranked a group of students for leadership ability and had another teacher rank the same children on attractiveness.

   a. What kind of scales are involved in this problem?
   b. What correlation coefficient is appropriate for this type of data?
   c. Using the following data and table as an aid, what is the value for r_s for this data?

Student	Leadership rank	Attractiveness rank	D²
A	1	2	
B	2	1	
C	3	3	
D	4	7	
E	5	4	
F	6	5	
G	7	6	

5a. Calculate the Pearson r for the following set of data:

X	1	2	3	4
Y	10	13	14	16

5b. Calculate the Pearson r for the following set of data:

X	1	2	3	4
Y	16	14	13	10

6. Given the following data, what are the values of b_Y and a_Y? (You may have to look back to the last chapter for a little reminder.)

$$s_X = 12 \quad s_Y = 14 \quad \sum X = 57 \quad \sum Y = 83 \quad N = 10 \quad r = .843$$

Answers: 1. 0.95 **2.** .75 or 75% (r^2) **3a.** 0.79 **3b.** .63 or 63% (r^2) **4a.** ordinal **4b.** Spearman rho **4c.** 0.75 **5a.** 0.98 **5b.** -.98 **6.** $b_Y = 0.984$ $a_Y = 2.691$

TRUE-FALSE QUESTIONS

T F 1. A correlation coefficient expresses the direction but not the magnitude of a relationship.

T F 2. Both Pearson r and Spearman rho can range from -1.00 to +1.00.

T F 3. The value of r obtained by calculating the correlation between X and Y is the same as the correlation between Y and X.

T F   4. If X and Y are plotted as standard (z) scores, then $r = b_Y$.

T F   5. If scores are z scores and if the slope equals 1 then z_X will always equal z_Y.

T F   6. One reason for calculating r from z scores is to make r independent of units and scaling.

T F   7. The coefficient of determination equals the proportion of variability accounted for by the relationship between the variables.

T F   8. The coefficient of determination equals $\sqrt{r}$.

T F   9. If $s_Y = s_X$ then $r = b_Y$.

T F 10. Since r is so widely used, it is appropriate to calculate r for nonlinear data.

T F 11. The formula for rho is actually just the formula for Pearson's r simplified to apply to lower order scaling.

T F 12. If the value of rho for a set of ordinal data equalled 0.68, the value of r for the same data would be 0.68.

T F 13. If the value of r on ratio scale raw data were 0.87 and the pairs of numbers were converted to ordinal data and r calculated for the ordinal data, r would equal 0.87.

T F 14. Restricting the range of either X or Y will generally lower the correlation between the variables.

T F 15. If one calculates r for a set of data and r equals 0.84, one can be certain that the relationship between the variables is not spurious.

T F 16. The correlation between two variables when N = 2 will always be perfect.

T F 17. The correlation coefficient when N = 2 is meaningless.

T F 18. If r = -1.00, the relationship is imperfect.

T F 19. If one calculates r for a set of numbers and then adds a constant to each value of one of the variables, the correlation will change.

T F 20. If the standard deviation of one of the variables equals zero, r cannot be calculated.

T F 21. The correlation coefficient r is a descriptive statistic.

T F 22. Using a second predictor vartiable always increases the accuracy of prediction.

SELF-QUIZ

1. Which of the following values of r represents the strongest degree of relationship between two variables?

   a.   0.55
   b.   0.00
   c.   0.78
   d.   -0.80

2. If X and Y are transformed into z scores, and the slope of the regression line is -0.80, what is the value of the correlation coefficient?

   a.   -0.80
   b.   0.80
   c.   0.40
   d.   -0.40

3. In order for the correlation coefficient to be negative, which of the following must be true?

   a. $\Sigma\,XY > (\Sigma\,X)(\Sigma\,Y)/N$
   b. $\Sigma\,XY < (\Sigma\,X)(\Sigma\,Y)/N$
   c. $\Sigma\,XY = (\Sigma\,X)(\Sigma\,Y)/N$
   d. $\Sigma\,XY$ must be zero

4. If two variables are ratio scaled and the relationship is linear, what type of correlation coefficient is most appropriate?

   a.   Pearson
   b.   Spearman
   c.   eta
   d.   phi

5. Correlation means causation.

   a.   true
   b.   false

6. Causation means correlation.

    a. true
    b. false

7. Pearson r can be properly used on which of the following type(s) of relationships?

    a. linear
    b. curvilinear
    c. exponential
    d. all of the above

8. If one takes a sample of pairs of points over a narrow range of X or Y scores, what effect might this have on the value of r?

    a. inflate r
    b. have no effect on r
    c. reduce r
    d. cannot be determined

9. You have conducted a brilliant study which correlates IQ score with income and find a value of r = .75. At the end of the study you find out all the IQ scores were scored 10 points too high. What will the value of r be with the corrected data?

    a. r will be increased
    b. r will be decreased
    c. r will remain the same
    d. cannot be determined

10. If z_X equals z_Y for each pair of points, r will equal _____.

    a.   0.00
    b. -1.00
    c.   1.00
    d.   0.50

11. If one calculates r for raw scores, and then calculates r on the z scores of the same data, the value of r will _____.

    a. stay the same
    b. decrease
    c. increase
    d. equal 1.00

12. You have noticed that as people eat more ice cream they also have darker sun tans. From this observation, you conclude _____.

    a. eating ice cream causes people to tan darker
    b. when one's skin tans it causes an urge to eat ice cream
    c. the results were spurious
    d. perhaps a third variable is responsible for the correlation
    e. all of the above are possible

13. If $s_Y = s_X = 1$ and the value of $b_Y = .6$, what will the value of r be?

    a. 0.36
    b. 0.60
    c. 1.00
    d. 0.00

14. If 49% of the total variability of Y is accounted for by X, what is the value of r?

    a. 0.49
    b. 0.51
    c. 0.70
    d. 0.30

15. What is the value of r for the following relationship between height and weight?

Height	60	64	65	68
Weight	103	122	137	132

    a. 0.87
    b. 0.76
    c. 0.93
    d. 0.56

16. For the following X and Y scores, how much of the variability of Y is accounted for by knowledge of X? Assume a linear relationship.

X	20	15	6	10
Y	6	5	4	0

    a. 68%
    b. 34%
    c. 58%
    d. 27%

17. What is the value of the Spearman rank order correlation coefficient (rho) for the following pairs of ranks?

Rank on A	Rank on B
1	3
2	1
3	4
4	2
5	5

a. 0.40
b. 0.50
c. 0.60
d. 0.70

18. In order to properly use rho, the variables must be of at least _____ scaling.

a. nominal
b. ordinal
c. interval
d. ratio

19. When using more than one predictor variable, _____ tells us the proportion of variance accounted for by the predictor variables.

a. r
b. SS_X
c. SS_Y
d. R^2

Answers: **1.** d **2.** a **3.** b **4.** a **5.** b **6.** a **7.** a **8.** c **9.** c **10.** c **11.** a **12.** e **13.** b **14.** c **15.** a **16.** b **17.** b **18.** b **19.** d.

8 RANDOM SAMPLING AND PROBABILITY

CHAPTER OUTLINE

I. Inferential Statistics. Uses the sample scores to make a statement about a characteristic of the population.

    A. <u>Hypothesis testing.</u> Data collected in an experiment in an attempt to validate some hypothesis involving a population.

    B. <u>Parameter estimation.</u> The experimenter is interested in determining magnitude of a population characteristic; e.g., population mean.

II. Methodology of Inferential Statistics

    A. <u>Random sampling.</u> A random sample is defined as a sample which has been selected from the population by a process which assures that each possible sample of a given size has an equal chance of being selected, <u>and</u> all the members of the population have an equal chance of being selected into the sample.

    B. Reasons for random sampling.

        1. Required in order to apply laws of probability to sample.

        2. It helps assure that the sample is representative of the population.

C. Use of random number table is one method of assuring random sampling.

D. Types of sampling.

1. Sampling with replacement. A method of sampling in which each member of the population selected for the sample is returned to the population before the next member is selected.

2. Sampling without replacement. A method of sampling in which the members of the sample are not returned to the population prior to selecting subsequent members.

III. Probability

A. Classical or *a priori* view of probability. That which can be deduced from reason alone. No actual data required.

$$p\ (A) = \frac{\text{the number of events classifiable as A}}{\text{the total number of possible events}}$$

where p(A) is read "The probability of occurrence of event A."

B. Empirical or *a posteriori* view of probability. Meaning after-the-fact or after some data has been collected.

$$p\ (A) = \frac{\text{the number of times A has occurred}}{\text{the total number of occurrences}}$$

C. Probability values.

1. Range from 0.00 to 1.00 (i.e., from an event is certain not to occur to an event is certain to occur).

2. Generally expressed as fraction or decimal.

D. Computing probability.

1. Addition rule. This rule is concerned with determining the probability of occurrence of any of several possible events. Rule for two events:

The probability of occurrence of A <u>or</u> B is equal to the probability of occurrence of A plus the probability of occurrence of B minus the probability of occurrence of both A and B, or

$$p(A \ or \ B) = p(A) + p(B) - p(A \ and \ B)$$

2. Multiplication rule. This rule is concerned with the joint or successive occurrence of several events. Rule for two events:
   The probability of occurrence of <u>both</u> A and B is equal to the probability of occurrence of A times the probability of occurrence of B given A has occurred, or

$$p(A \ and \ B) = p(A)p(B|A)$$

   Both rules can be extended to more than two events.

E. Additional concepts.

   1. Mutually exclusive events. Two events are mutually exclusive if they both cannot occur together (e.g., a coin coming up both heads and tails on one flip).

   2. Exhaustive events. A set of events is exhaustive if the set includes all of the possible outcomes (e.g., for the flip of a coin, head and tail is the exhaustive set of possible outcomes).

   3. Mutually exclusive and exhaustive events. When a set of events is both mutually exclusive and exhaustive the sum of the individual probabilities of each event in the set must equal 1.00. Thus, under these conditions:

$$p(A) + p(B) + p(C) +\cdots+ p(Z) = 1.00$$

   4. When there are two events and they are mutually exclusive and exhaustive, then:

$$P + Q = 1.00$$

IV. More on the Multiplication Rule

A. <u>Mutually exclusive events.</u> If A and B are mutually exclusive, then:

$$p(A \text{ and } B) = 0$$

B. <u>Independent events.</u> Two events are independent if the occurrence of one has no effect on the probability of occurrence of the other. In this case the rule becomes:

$$p(A \text{ and } B) = p(A)p(B|A) = p(A)p(B)$$

C. <u>Dependent events.</u> Two events are dependent if the probability of occurrence of B is affected by the occurrence of A. In this case the rule becomes:

$$p(A \text{ and } B) = p(A)p(B|A)$$

D. Multiplication and addition rules can be combined to solve problems.

V. Probability and Normally Distributed Continuous Variables

A. Probability of A with a continuous variable can be expressed:

$$p(A) = \frac{\text{the area under the curve corresponding to A}}{\text{the total area under the curve}}$$

B. Solution:

1. Convert raw score to its transformed value (its standard score).

2. Look up area in Table A.

CONCEPT REVIEW

A basic aim of (1) _____ statistics is to use (2) _____ scores to make a statement about a characteristic of the population. The characteristic may be either (3) _____ testing or (4) _____ (5) _____. Essential to the methodology of

(1) inferential
(2) sample

(3) hypothesis

(4) parameter
(5) estimation

inferential statistics are(6) _____ sampling and (7) _____.

(6) random
(7) probability

A random sample is one which has been selected from the

(8) _____ by a process which assures that each possible

(8) population

(9) _____ of a given size has an (10) _____ chance of being

(9) sample
(10) equal

selected and that all the (11) _____ of the population have an

(11) members

(12) _____ chance of being selected into the sample.

(12) equal

Sampling should be random for two reasons. First, to apply

the (13) _____ of (14) _____ sampling must be random.

(13) laws
(14) probability

Second, in order to (15) _____ it is necessary that the sample

(15) generalize

be (16) _____ of the population. If a sample is (17) _____

(16) representative

it may lead to an erroneous conclusion. One way to help assure

(17) biased

random sampling is by the use of a table of (18) _____

(18) random

(19) _____.

(19) numbers

In probability theory there can be two kinds of sampling. One

method is called sampling with (20) _____. In this case

(20) replacement

each member of the population selected for the sample is

(21) _____ to the population (22) _____ the next member is .

(21) returned
(22) before

selected. The other method is sampling (23) _____ replacement.

(23) without

In this case the members of the sample are (24) _____

(24) not

(25) _____ to the population (26) _____ selecting subsequent

(25) returned
(26) before

members.

Probability can be approached in two ways, from an

(27) _____ (28) _____ approach, also called the

(27) *a*
(28) *priori*

(29) _____ viewpoint or from the (30) _____ (31) _____

(29) classical
(30) *a*

or empirical approach. The symbol (32) _____

means the probability of the occurrence of event A. From the

classical viewpoint, probability is defined:

(31) *posteriori*
(32) p(A)

$$p(A) = \frac{\text{the (33) \_\_\_\_ of events (34) \_\_\_\_ as A}}{\text{the (35) \_\_\_\_ number of (36) \_\_\_\_ events}}$$

(33) number
(34) classifiable
(35) total
(36) possible

From the empirical perspective, probability can be defined:

$$p(A) = \frac{\text{the number of (37) \_\_\_\_ A has occurred}}{\text{the   total number of (38) \_\_\_\_}}$$

(37) times

(38) occurrences

In this latter approach, actual (39) _____ must be collected. If

enough data is collected the two probabilities will (40) _____

each other if chance alone is involved.

(39) data

(40) equal

Since probability is a (41) _____, the values can range

from (42) _____ to (43) _____. If the probability equals

0.00, then the event is certain (44) _____ to occur. If the prob-

ability equals (45) _____, the event is certain to occur.

(41) proportion

(42) 0.00
(43) 1.00
(44) not

(45) 1.00

The addition rule states that the probability of occurrence of

A (46) _____ B is equal to probability of occurrence of A

(47) _____ the probability of occurrence of B (48) _____ the

probability of occurrence of (49) _____ A (50) _____ B. In

equation form:

(46) or

(47) plus
(48) minus
(49) both
(50) and

$$p(A \text{ or } B) = (51) \_\_\_\_ + (52) \_\_\_\_ - (53) \_\_\_\_$$

(51) p(A)
(52) p(B)
(53) p(A and B)

This rule is generally used when events are (54) _____ ex-

(54) mutually

clusive. This means that the occurrence of one event (55) _____ (55) precludes
the occurrence of the other. When two events are mutually exclusive
then the addition rule becomes:

$$p(A \text{ or } B) = (56) \text{\_\_\_\_\_} + (57) \text{\_\_\_\_\_}$$

(56) p(A)
(57) p(B)

This is true because p(A and B) = (58) _____. A set of events

(58) 0

is (59) _____ if the set includes all of the possible events.

(59) exhaustive

When events are both mutually exclusive and exhaustive, the

(60) _____ of the individual probabilities of each event in

(60) sum

the set must equal (61) _____. If we call one event P and

(61) 1.00

the other Q and if P and Q are both mutually exclusive and exhaustive then this can be expressed as

$$(62) \text{\_\_\_\_} \_ + (63) \text{_____} = (64) \text{_____}$$

(62) P
(63) Q
(64) 1.00

The multiplication rule states the probability of the occurrence of A (65) _____ B is equal to the probability of occurr-

(65) and

ence of A (66) _____ the probability of B given A has occurred.

(66) times

In equation form:

$$p(A \text{ and } B) = (67) \text{_____}$$

(67) p(A)p(B|A)

The term (68) _____ means the probability of occurrence of B

(68) p(B|A)

given (69) _____ has occurred.

(69) A

When A and B are mutually (70) _____ then p(A and B)

(70) exclusive

= (71) _____.

(71) 0

Two events are considered (72) _____ if the occurrence   (72) independent

of one has (73) _____ effect on the probability of the other.   (73) no

In this case p(BIA) = (74) _____. The multiplication rule   (74) p(B)

then becomes:

$$p(A \ and \ B) = (75) \ \underline{\hspace{1.5cm}}$$ (75) p(A)p(B)

The original formula used earlier, namely, p(A and B) =

p(A)p(BIA) is used when A and B are (76) _____. This   (76) dependent

means the probability of B is (77) _____ by the occurrence   (77) affected

of A. One common example of a dependent event occurs when

one samples (78) _____ replacement. For some problems   (78) without

we can use both the multiplication and the addition rules together.

So far we have been discussing (79) _____ variables.   (79) discrete

When a variable is (80) _____ probability can be defined as   (80) continuous

$$p(A) = \frac{the \ (81) \ \underline{\hspace{0.8cm}} \ under \ the \ curve \ corresponding \ to \ A}{the \ (82) \ \underline{\hspace{0.8cm}} \ area \ under \ the \ curve}$$ (81) area
          (82) total

Often these variables are normally distributed. In that case we

can find the area under the curve by converting the (83) _____   (83) raw

score to its (84) _____ value or (85) _____ score and then   (84) z
          (85) z

looking up the area in Table A. We recall that this transformation

is made by

$$z = \frac{(86) \ \underline{\hspace{1cm}} - (87) \ \underline{\hspace{1cm}}}{(88) \ \underline{\hspace{1cm}}}$$ (86) X
          (87) μ
          (88) σ

EXERCISES

1. A gambler gives you a coin which he tells you is a "trick" coin. That is, the probability of getting a head does not equal the probability of getting a tail. To use the coin to your advantage you need to know the probability of obtaining a head; i.e., p(H).

    a. How can you determine p(H)?
    b. Assuming you tossed the coin 5000 times and got 1900 tails, what would you estimate p(H) to be?
    c. What method of determining probability is this called?

2. In a class of 30 children who cannot swim, the instructor randomly selects a child and then teaches the child to swim.

    a. What was the probability that any one particular child would have been selected?
    b. What is the probability that if the process is repeated a second time that a particular child will be selected?
    c. What type of sampling technique is this?

3. The local psychological association reports that in a particular community there are 36 analytic psychologists, 57 behaviorists, 92 cognitive psychologists, 17 Gestalt psychologists and 42 of unknown theoretical background.

    a. If you randomly select a psychologist from a list containing all the above psychologists, what is the probability you will select either a known behaviorist or a known analytic psychologist? Assume sampling with replacement for all parts of this problem.
    b. What is the probability that you would select a psychologist of unknown background?
    c. What is the probability that you will select a psychologist with a known background?
    d. If you draw two psychologists, what is the probability that you will draw a known cognitive psychologist and a known Gestalt psychologist in that order?

4. In a store which sells health foods, there are 16 different types of flour. Four of the flours are white and the rest are brown. Assume sampling with replacement.

    a. Assuming random selection, what is the probability of selecting a brown or white flour?
    b. What is the probability of selecting a white and then a brown flour?
    c. What is the probability of selecting a brown and then a white flour?
    d. What is the probability of selecting three brown flours in a row?

5. Consider an ordinary deck of playing cards. What is the probability for the following:

    a. Drawing a red card or a king?

    b. Drawing an ace, king, queen, jack, and 10 of the same suit in that order without replacement?

    c. Drawing a face card (K, Q or J)?

    d. Drawing 4 threes in four draws without replacement?

    e. Drawing 4 threes in four draws with replacement?

6. A cattle breeder has 100 black cows of breed X, 50 brown cows of breed Y and 22 white cows of breed Y.

    a. What is the probability of randomly selecting a white cow and having it be a member of breed X?

    b. What kind of events are the events described in part a?

    c. What is the probability of randomly selecting a cow and having it be brown or of breed X?

    d. What is the probability of selecting at random and with replacement, first a brown cow, then a white cow, and finally a cow that is either of breed X or white?

7. Prior to taking an exam of 5 true or false items, assume (as ludicrous as this seems) that you did not have a chance to study. Assume further that you just have to guess on each item with the resulting probability of getting any item correct is .50. What is the probability for each of the following outcomes?

    a. That you get all 5 items correct?

    b. That you miss all 5 items?

    c. That you miss item 2?

    d. That you correctly answer the even number items?

    e. That you get both the first 2 items correct or fifth item correct?

8. If the average survival of a red cell in the human body is 120 days with a standard deviation of 8 days, what is the probability for each of the following events? Assume a normal distribution.

    a. A red cell surviving more than 134 days?

    b. A red cell that survives less than 110 days?

    c. A red cell surviving between 108 and 122 days?

9. At the local race track 10 races are run each day with 10 different horses in each race. You bet on a horse in each race but select the horse by using a random number table.

    a. What is the probability you will choose a winner the first three races?

    b. What is the probability you will choose a winner in race 6?

    c. Does the fact that horse number 5 wins the third race affect the probability that any other number 5 will win during the day?

10. In an experiment on parapsychology a scientist wants to investigate how well a subject can correctly guess what symbol the experimenter is thinking of. The experimenter chooses from 4 different possible symbols. To avoid bias the experimenter uses a random number table to select a symbol to concentrate on for

each trial. Assuming the experiment is not biased what are the probabilities for the following outcomes if chance alone is operating?

a. A subject identifying the first 3 symbols in a row correctly?
b. A subject missing the first 5 in a row?
c. The subject guessing the nth symbol correctly

11. What is the probability that you can correctly guess the spelling of a three letter sequence that you do not know, if the first letter is different from the second and third letters but the second and third letters may or may not be the same?

12. What is the probability that you can correctly guess a three letter sequence if the first letter is a consonant and the second and third letters are the same vowel (a, e, i, o, u, or y)?

13. A normally distributed continuous variable has a value of $\mu = 60$ and $\sigma = 14$. If one draws a score from the distribution, what is the probability that it will be:

a. between 38 and 60?
b. between 68 and 74?
c. less than 60?
d. less than 59?
e. greater than 40.1?

14. In a box there are 10 slips of paper and each slip of paper has a number from 0 to 9 on it so that all the numbers appear once and only once.

a. What is the probability that if you draw 3 slips of paper in a row with replacement the digits would be 1, 2, 3?
b. If you draw 4 slips one-at-a-time with replacement, what is the probability that the number resulting from the four digits will be even?
c. What is the probability that in one draw you will draw a number less than 4 or greater than 7?

15. What is the probability of throwing a pair of dice and having the sum of the dice equal 7 or 11?

16. Approximately how many times out of 100 could you throw a pair of dice 5 times and not come up with a total of 7 or 11?

Answers: **1a.** Take the coin and flip it a large number of times and observe the occurrence of the event of interest, namely a head occurring and then apply the formula: $p(H) = $ (the number of heads)/(the number of flips) **1b.** 3100/5000 = .6200 **1c.** Empirical or *a posteriori* method **2a.** 1/30 = .0333 **2b.** 1/29 = .0345 **2c.** sampling without replacement since once a child has learned to swim s/he cannot be returned to the population to be drawn again **3a.** .3811 **3b.** .1721 **3c.** 1 - .1721 = .8279

3d. .0263   **4a.** 1.00   **4b.** .1875   **4c.** .1875   **4d.** .4219   **5a.** .5385   **5b.** 4/52 x 1/51 x 1/50 x 1/49 x 1/48 = .00000001283   **5c.** 12/52 = .2308   **5d.** 4/52 x 3/51 x 2/50 x 1/49 = .000003694   **5e.** $(4/52)^4$ = .00003501   **6a.** 0 **6b.** mutually exclusive   **6c.** .8721   **6d.** 50/172 x 22/172 x (100/172 + 22/172) = .0264   **7a.** $.5^5$ = .0312   **7b.** $.5^5$ = .0312   **7c.** .5000   **7d.** $.5^2$ = .2500   **7e.** .5 x .5 + .5 = .7500   **8a.** .0401   **8b.** .1056   **8c.** .5319   **9a.** $1/10^3$ = .0010   **9b.** .1   **9c.** no   **10a.** $(1/4)^3$ = .0156   **10b.** $(3/4)^5$ = .2373   **10c.** 1/4 = .2500   **11.** 1/26 x 1/25 x 1/25 = .00006154   **12.** 1/20 x 1/6 x 1/1 = .008333   **13a.** .4418   **13b.** .1256   **13c.** .5000   **13d.** .4721   **13e.** .9222   **14a.** .001   **14b.** .50   **14c.** .6000   **15.** 8/36 = .2222   **16.** approximately 28 on the average (.2847 x 100).

TRUE-FALSE QUESTIONS

T F   1. Hypothesis testing is part of inferential statistics while parameter estimation is used in descriptive statistics.

T F   2. In order to generalize to the population a sample must be randomly selected.

T F   3. For a sample to be random, all the members must have an equal chance of being selected into the sample.

T F   4. To use a random number table properly one must begin on the top left hand column and read across.

T F   5. Sampling without replacement is always used in choosing subjects for an independent groups design experiment.

T F   6. An *a posteriori* approach to probability is never used because it is only an approximation of the true probability.

T F   7. Probability values range from -1.00 to +1.00.

T F   8. Events that occur only rarely have a probability equal to 0.00.

T F   9. Two events are considered mutually exclusive if the probability of one event does not influence the probability of a second event.

T F   10. When two events are dependent, then p(A and B) = p(A) + p(B).

T F   11. When events are mutually exclusive and exhaustive then the sum of the individual probabilities of each event in the set must equal 1.00.

T F   12. The addition rule is concerned with determining the probability of A or B while the multiplication rule is concerned with determining the probability A and B.

T F 13. When two events are mutually exclusive then p(A and B) must equal 1.00.

T F 14. Two events are independent if the occurrence of one has no effect on the probability of occurrence of the other.

T F 15. If A and B are independent then p(B|A) = p(A) or p(B).

T F 16. For dependent events, p(A and B) = p(A)p(B|A).

T F 17. The addition and multiplication rules can apply to any number of events.

T F 18. The multiplication and addition rules can be applied together in the same problem in order to calculate probabilities under some circumstances.

T F 19. There is no division rule for probability.

T F 20. To use probability and inference, data must be discrete.

T F 21. For continuous variables which are normally distributed, p(A) equals the area under the curve corresponding to A divided by the total area under the curve.

T F 22. To calculate the probability of drawing three sevens in a row from a deck of cards involves the use of the multiplication rule and equals 4/52 + 3/51 + 2/50.

Answers: 1. F **2.** T **3.** T **4.** F **5.** T **6.** F **7.** F **8.** F **9.** F **10.** F **11.** T **12.** T **13.** F **14.** T **15.** F **16.** T **17.** T **18.** T **19.** T **20.** F **21.** T **22.** F.

SELF-QUIZ

1. If a town of 7000 people has 4000 females in it, then the probability of randomly selecting 6 females in six draws (with replacement) equals _____.

    a. .0348
    b. .0571
    c. .5714
    d. .3429

2. If a stranger gives you a coin and you toss it 1,000,000 times and it lands on heads 600,000 times, what is p(Heads)I for that coin?

    a. .5000
    b. .6000
    c. .4000
    d. .0000

3. The probability of randomly selecting a face card (K, Q, or J) or a spade in one draw equals _____.

    a. .0192
    b. .0577
    c. .4808
    d. .4231

4. The probability of drawing an ace followed by a king (without replacement) equals _____.

    a. .0044
    b. .0060
    c. .0045
    d. .0965

5. The probability of throwing two ones with a pair of dice equals _____.

    a. .3600
    b. .1667
    c. .0278
    d. .3333

6. If p(A or B) = p(A) + p(B) then A and B must be _____.

    a. dependent
    b. mutually exclusive
    c. overlapping
    d. continuous

7. If P + Q = 1.00 then P and Q must be _____.

    a. mutually exclusive
    b. exhaustive
    c. random
    d. a and b

8. If $\mu = 35.2$ and $\sigma = 10$, then $p(X)$ for $X \leq 39$ equals _____. Assume random sampling.

   a. .3520
   b. .6200
   c. .1480
   d. .6480

9. If $p(A$ and $B) = 0$, then A and B must be _____.

   a. independent
   b. mutually exclusive
   c. exhaustive
   d. unbiased

10. If $p(A)p(B|A) = p(A)p(B)$, then A and B must be _____.

    a. independent
    b. mutually exclusive
    c. random
    d. exhaustive

11. If $p(A$ and $B) = p(A)p(B|A) \neq p(A)p(B)$, then A and B are _____.

    a. mutually exclusive
    b. random
    c. independent
    d. dependent

12. If $p(A) = 0.6$ and $p(B) = 0.5$, then $p(B|A)$ equals _____.

    a. .8333
    b. .3000
    c. .5000
    d. cannot be determined from the information given

13. If $\mu = 400$ and $\sigma = 100$ the probability of selecting at random a score less than or equal to 370 equals _____.

    a. .1179
    b. .6179
    c. .3821
    d. .8821

14. If you have 15 red socks (individual, not pairs), 24 green socks, 17 blue socks, and 100 black socks, what is the probability you will reach in the drawer and randomly select a pair of green socks? (Assume sampling without replacement.)

   a. .3022
   b. .3077
   c. .0228
   d. .0227

15. If the probability of drawing a member of a population is not equal for all members, then the sample is said to be _____.

   a. random
   b. independent
   c. exhaustive
   d. biased

16. The probability of rolling an even number or a one on a throw of a single die equals _____.

   a. .6667
   b. .5000
   c. .0834
   d. .3333

17. If events are mutually exclusive they cannot be _____.

   a. independent
   b. exhaustive
   c. related
   d. all the above

18. The probability of correctly guessing a two digit number is _____.

   a. .1000
   b. .0100
   c. .2000
   d. .5000

19. When events A and B are mutually exclusive but not exhaustive, p(A or B) equals _____.

   a. 0.50
   b. 0.00
   c. 1.00
   d. cannot be determined from the information given

20. The probability of correctly calling 4 tosses of an unbiased coin in a row equals
    _____.

    a. .0625
    b. .5000
    c. .1250
    d. .2658

Answers: 1. a   **2.** b   **3.** d   **4.** b   **5.** c   **6.** b   **7.** d   **8.** d   **9.** b   **10.** a
11. d   **12.** d   **13.** c   **14.** c   **15.** d   **16.** a   **17.** a   **18.** b   **19.** d   **20.** a.

9 | BINOMIAL DISTRIBUTION

CHAPTER OUTLINE

I. **Binomial Distribution**

    A. <u>Definition.</u> This is a probability distribution which results from a certain set of circumstances.

        1. Series of N trials.

        2. Only 2 possible outcomes on each trial.

        3. Outcomes are mutually exclusive.

        4. Each trial is independent of every other trial.

    B. Information obtained.

        1. Each possible outcome of the N trials.

        2. Probability of getting each of the possible outcomes.

    C. Example - "Flipping a Coin"

        1. Each flip is a trial.

2. Only 2 possible outcomes.

3. Mutually exclusive because only a head or tail can occur.

4. Independent; i.e., outcome of one flip doesn't affect any other flip.

II. Binomial Expansion. Avoids having to enumerate outcomes of binomial events.

A. <u>Formula:</u>   **$(P + Q)^N$**

where   P = probability of one of the two possible outcomes on a trial
Q = probability of the other possible outcome (1 - P)
N = number of trials

B. <u>Interpretation.</u> The letters in each term tell the kind of event and the exponents tell the number of that kind of event; e.g., P^2 means there are two P events.

C. <u>Binomial distribution.</u> The binomial distribution may be generated for any N, P and Q by using the binomial expansion. Note that:

1. This distribution is symmetrical with P = .50.

2. As N increases, the distribution approximates a normal curve.

D. <u>Binomial table.</u> This table shows the results for the binomial expansion solved for many values of N, P or Q, and no. of P or Q events. Table can be used with P or Q.

CONCEPT REVIEW

The binomial distribution is a (1) _____ distribution.

A probability distribution tells us each possible (2) _____

and the (3) _____ of getting each of the outcomes. (4) _____

conditions are necessary to utilize the binomial distribution. They

are:

1. There is a series of (5) _____ (6) _____;

2. Each trial has only (7) _____ possible outcomes;

3. The outcomes are (8) _____ (9) _____;

(1) probability

(2) outcome

(3) probability
(4) Four

(5) N
(6) trials
(7) two

(8) mutually
(9) exclusive

4. There is (10) _____ between the outcomes of each trial. (10) independence

Tossing a coin is an example of a binomial event. If you toss

two coins there are four possible outcomes. They are (11) HH

(11) _____, (12) _____, (13) _____, or (14) _____. (12) HT
(13) TH

The probability of getting two heads is (15) _____. The (14) TT
(15) 1/4 or .25

probability of getting two tails is (16) _____. From this infor- (16) 1/4 or .25

mation we can construct the probability distribution. For tossing

two coins the following table results:

Outcome	f	Probability
2 heads	(17) _____	(18) _____
1 head, 1 tail	(19) _____	(20) _____
2 tails	(21) _____	(22) _____

(17) 1
(18) .25
(19) 2
(20) .50
(21) 1
(22) .25

From this table we can determine each of the

possible (23) _____ and the (24) _____ of obtaining (23) outcomes
(24) probabilities

those outcomes. For N's > 2 we can continue this process

of (25) _____ to determine the outcomes and their proba- (25) enumeration

bilities. Another way of obtaining the binomial distribution for a

given size N is to use the (26) _____ (27) _____. The (26) binomial
(27) expansion

binomial expansion is given by ((28) _____ + (29) _____)N. (28) P
(29) Q

P equals the (30) _____ of any one of the two possible (30) probability

outcomes on a trial, while (31) _____ equals the probability (31) Q

of the other possible outcome. N is the (32) _____ of. (32) number

(33) _____. (33) trials

For the example of two coins, the value P = Q = (34) _____. (34) .50

For two tosses of the coin the expansion equals P^2 +

2(35) _____ + Q^2. The terms (36) _____, (37) _____,

and (38) _____ represent all the possible outcomes of

flipping the (39) _____ coins once. The term P^2 tells us this

outcome consists of 2 P events or 2 (40) _____.

The 2PQ term tells us that the outcome consists of (41) _____

head and (42) _____ tail. The Q^2 tells us the outcome consists

of (43) _____ heads. If no heads occurred then the outcome

must be two (44) _____. If we wanted to know the probability

of getting one head and one tail we would solve the (45) _____

term. Substituting the value of (46) _____ for P and Q, the

expression becomes (47) _____ ((48) _____) ((49) _____),

which equals (50) _____.

(35)	PQ
(36)	P^2
(37)	2PQ
(38)	Q^2
(39)	two
(40)	heads
(41)	one
(42)	one
(43)	0
(44)	tails
(45)	2PQ
(46)	.50
(47)	2
(48)	.50
(49)	.50
(50)	.50

    Fortunately, the results of the binomial expansion are summa-

rized in Table (51) _____. (52) _____ is given in the first

column and the possible (53) _____ are given in the second

column. The rest of the columns contain probability values

for various values of (54) _____.

(51)	B
(52)	N
(53)	outcomes
(54)	P or Q

    For example, if one wanted to know what the probability of

getting 6 heads in one toss of 10 unbiased coins, one would go to

table entry:

N	No. of P Events	P
(55) _____	(56) _____	(57) _____

(55)	10
(56)	6
(57)	.50

where one finds the probability equals (58) _____. If one (58) .2051

wants to find the probability of obtaining 6 or more heads in

one toss of the 10 coins, one simply (59) _____ the prob- (59) adds

abilities found in the table for (60) _____, (61) _____, (60) 6
(61) 7
(62) _____, (63) _____, and (64) _____ P events. (62) 8
(63) 9

    Table B can also be used for values of (65) _____ ≠ .50. (64) 10
(65) P

If P = .30 and N = 12, the p(8 P events) = (66) _____ (66) .0078

If P > .50, the problem is solved by using (67) _____. For (67) Q

example, if P = .85 and N = 7, to find p(5 P events), the table is

entered under:

N	No. of Q Events	Q
(68) _____	(69) _____	(70) _____

(68) 7
(69) 2
(70) .15

Thus, p(5 P events) = (71) _____. (71) .2097

EXERCISES

1. What is the probability of tossing 8 unbiased coins once and obtaining at least 5 heads?

2. What is the probability of tossing 6 unbiased coins once and getting 6 heads? What is the probability of getting 6 tails? What is the probability of obtaining either 6 heads or 6 tails?

3. What is the probability of rolling a die and getting an even number exactly 5 times out of 9 rolls?

4. Develop the binomial distribution for the experiment of tossing 3 unbiased coins once. Use the following table to help you set up the problem.

Outcome	f	Probability
0 heads	_ _ _ _ _	_ _ _ _ _
1 head	_ _ _ _ _	_ _ _ _ _
2 heads	_ _ _ _ _	_ _ _ _ _
3 heads	_ _ _ _ _	_ _ _ _ _

5. If you were at a race track and bet on 9 races in a day, each with 10 horses entered, what is the probability of winning exactly 2 races if you were picking your winners by guessing alone?

6. If you flipped 8 coins once what is the probability of getting results more extreme than 6 heads? Assume the probability of a head with each coin = .75.

7. If you weighted a coin such that the probability of obtaining a head on any one flip was .3, what is the probability of getting 5 or 6 heads out of 6 flips? What is the probability of getting 0 or 1 heads out of 6 flips?

8. A coin is said to be biased if $p(H) \neq p(T) \neq .50$. If unfair coins are biased such that $p(H) = .35$, what is the probability of getting 5 or more heads if 10 coins were tossed once?

9. What is the probability of obtaining 2 or fewer heads if $p(H) = .2$ and 7 coins were tossed once?

10. If $p(H) = .10$ and 12 coins were flipped once, what are the probabilities for the following outcomes?

    a. Obtaining exactly 3 heads?
    b. Obtaining 0 heads?
    c. Obtaining 4 or fewer heads?
    d. Obtaining results as extreme or more extreme than 4 heads?

11. If $p(H) = .75$ what is the probability that if 9 coins were tossed once that one would get 6 or more heads?

12. If $p(H) = 1.00$ and N coins were tossed, how many coins would come up tails?

13. If $p(H) = .90$ and 8 coins were tossed once, what is the probability of getting results as extreme or more extreme than 7 heads?

14. If $p(H) = .80$ and 9 coins were tossed once, what is the probability of getting results as extreme or more extreme than 3 heads?

15. If the probability of a defective chair = 0.85 and you randomly sample 20 chairs from 10,000 chairs, what is the probability that at least 19 chairs will be defective?

16. A particular industry maintains that from the outset within it women have had the same chance of being hired as men. If this is so, of 10,000 companies within the industry that employ 14 individuals, how many companies would you expect to have 13 or more male employees?

Answers: 1. .3633   **2.** .0156, .0156, .0312   **3.** .2461

4. binomial distribution table.

$\underline{f}$	Probability
1	.1250
3	.3750
3	.3750
1	.1250

5. .1722   **6.** .3671   **7.** .0109, .4201   **8.** .2485   **9.** .8520   **10a.** .0852   **10b.** .2824   **10c.** .9956   **10d.** .9956   **11.** .8343   **12.** 0   **13.** .8131   **14.** .9175   **15.** .1756   **16.** 10.

TRUE-FALSE QUESTIONS

T F   1. In order to correctly apply the binomial distribution, one of the conditions that must be met is that there are only two possible outcomes.

T F   2. To apply the binomial distribution, three of the conditions which must be met are that there is a series of N trials where the outcomes are mutually exclusive and there is independence between trials.

T F   3. A valid example of a situation where one can apply the binomial distribution is in determining the probability of rolling a 6 or a 5 with the toss of a pair of dice.

T F   4. One can appropriately apply the binomial distribution if P = .37 and Q = .63.

T F   5. One can appropriately apply the binomial distribution if P = .5 and Q = .3 in a problem.

T F   6. To apply the binomial distribution properly (P + Q) must equal 1.00.

T F   7. $(P + Q)^2 = P^2 + Q^2$ in the binomial expansion.

T F   8. If P ≠ Q, the binomial distribution will still be normally distributed.

T F   9. The binomial distribution applies equally well to discrete and continuous variables.

T F 10. In $(P + Q)^7$ the term Q^7 indicates 0 P events.

T F 11. The probability of 6 P events when N = 10 and P = .7 is the same as 4 Q events if Q = .3 and N = 10.

T F 12. If N > 20, the binomial distribution cannot be used.

T F 13. If N = 18 the probability of obtaining a result as extreme or more extreme than 16 P events equals p(16) + p(17) + p(18).

T F 14. For biased coins, the probability of getting 5 heads out of a toss of 5 coins equals the probability of getting 5 tails out of a toss of 5 coins.

T F 15. One can look at picking a winner or not picking a winner in a series of races at the track as fitting the requirements of the binomial distribution (assuming that each horse has an equal chance of winning a particular race and there are the same number of horses in each race).

T F 16. The probability of getting a result as extreme or more extreme than 5 heads out of a toss of 7 unbiased coins is .4532.

Answers: 1. T   **2.** T   **3.** F   **4.** T   **5.** F   **6.** T   **7.** F   **8.** F   **9.** F   **10.** T
11. T   **12.** F   **13.** F   **14.** F   **15.** T   **16.** T.

SELF-QUIZ

1. Solving the binomial expansion to 4 decimal places will give more accurate answers than Table B.

    a. True
    b. False

2. If an event has 3 possible outcomes, then one cannot use the binomial distribution in analyzing the results.

    a. True
    b. False

For problems 3 and 4 consider the binomial expansion for 6 events:

$$P^6 + 6P^5Q + 15P^4Q^2 + 20P^3Q^3 + 15P^2Q^4 + 6PQ^5 + Q^6$$

3. What term would you use in evaluating the probability of obtaining exactly 2 heads as the result of flipping an unbiased coin 6 times?

    a. P^6
    b. $20P^3Q^3$
    c. $15P^2Q^4$
    d. the entire expression

4. What term(s) would you use to evaluate the probability of obtaining 1 or fewer heads from 6 flips of an unbiased coin?

    a. $6PQ^5$
    b. $6PQ^5 + Q^6$
    c. Q^6
    d. $20P^3Q^3$

5. If $Q = .30$ then P equals _____.

    a. .30
    b. .90
    c. .70
    d. .10

6. What is the value of $6P^5Q$ if $P = .20$?

    a. .0003
    b. .1600
    c. .0469
    d. .0015

7. In order to use the binomial distribution, which of the following conditions are necessary?

    a. a series of N trials with only 2 possible outcomes
    b. outcomes are mutually exclusive
    c. there must be independence between trials
    d. all of the above

8. If $P = .40$ for the probability of getting a head on any one flip of the coin, then _____.

    a. the odds of getting 4 out of 4 heads equals the odds of getting 4 out of 4 tails
    b. the odds of getting a head on any one toss is greater than the odds of getting a tail
    c. the odds of getting a tail is greater than the odds of getting a head
    d. cannot be determined

9. In the binomial expansion of $(P + Q)^{10}$, the last term of the expansion will be _____.

    a. Q^{10}
    b. P^{10}
    c. $P^{10} + Q^{10}$
    d. depends on the values of P and Q

10. The sum of all terms in any binomial expansion will equal _____.

    a. 1.0000
    b. 0.5000
    c. depends on the values of P and Q
    d. 0.0000

11. P^0 equals _____.

    a. 1.00
    b. 0.00
    c. 0.50
    d. Q

12. If one flips 5 unbiased coins once, what is the probability of getting exactly 3 heads?

    a. .6000
    b. .3125
    c. .4687
    d. .0312

13. What is the probability that one can call the flip of a coin correctly at least 6 out of 7 times assuming that the coin is fair?

    a. .0078
    b. .0547
    c. .5000
    d. .0625
    e. .1250

14. What is the probability of 4 heads turning up out of four tosses of the coin if the probability of any one head is $P = .40$?

    a. .0625
    b. .1296
    c. .0256
    c. .3456

15. In one flip of 10 unbiased coins, what is the probability of getting 8 or more heads?

    a. .0547
    b. .0016
    c. .0439
    d. .0010

16. In one flip of 10 unbiased coins, what is the probability of getting a result as extreme or more extreme than 8 heads?

    a. .0547
    b. .1094
    c. .0020
    d. 1.0000

17. If one of the terms in a binomial expansion were $210P^6Q^4$, what is the value of N?

    a. 210
    b. 6
    c. 4
    d. 10

18. What is the probability of obtaining exactly 5 P events if the appropriate term of the binomial expansion to evaluate were $6P^5Q^1$ and P = .42?

    a. .0784
    b. .0455
    c. .4616
    d. .0076

19. The probability of obtaining a result as extreme or more extreme than 4 heads if 11 coins were tossed and p(H) = .35 is _____.

    a. .6683
    b. .2745
    c. .5490
    d. .7184

20. A football player is practicing making field goals from the 30-yard line. If the probability of his kicking a field goal is 0.75, what is the probability he will kick at least 12 field goals in the next 15 tries? Assume independence between tries.

    a. .0000
    b. .2252
    c. .4613
    d. .6481

Answers: 1. b **2.** a **3.** c **4.** b **5.** c **6.** d **7.** d **8.** c **9.** a **10.** a **11.** a **12.** b **13.** d **14.** c **15.** a **16.** b **17.** d **18.** b **19.** a **20.** c.

10 INTRODUCTION TO HYPOTHESIS TESTING USING THE SIGN TEST

CHAPTER OUTLINE

1. The Experiment

A. <u>Purpose.</u> An experiment allows the scientist to test the influence of the independent variable upon the dependent variable while controlling for the influence of other variables.

B. <u>Conclusions.</u> In making a conclusion based upon an experiment, one has to answer the question, "How reasonable are these results if chance alone were responsible for the results?" If the results are not likely to be due to chance, then the results are attributed to the experimental manipulation.

C. <u>Repeated measures design.</u> This is one of several commonly used designs. It is also called the replicated measures design or correlated groups design. The essential feature is that subjects are paired prior to conducting the experiment and the difference between paired scores are analyzed.

D. <u>Hypothesis testing.</u>

1. Alternative hypothesis (H_1). The alternative hypothesis claims that the difference in results between conditions is due to the independent variable.

   a. Directional hypothesis. The hypothesis specifies the direction of the effect of the independent variable; e.g., marijuana slows reaction time.

143

    b. Nondirectional. The hypothesis states that the independent variable has an effect but the direction of the effect is not stated.

2. Null hypothesis (H_0). The null hypothesis is the logical counterpart of the alternative hypothesis. For a nondirectional alternative hypothesis, the null hypothesis states that the independent variable has no effect on the dependent variable. For a directional alternative hypothesis, the null hypothesis states that the independent variable has no effect in the direction predicted by H_1.

3. Mutually exclusive and exhaustive. The alternative hypothesis and the null hypothesis are mutually exclusive and exhaustive. If one is true the other cannot be. If one is false the other must be true. We always analyze the null hypothesis and try to show that it is false. If we show the null hypothesis is false, then we can accept the alternative hypothesis as true.

4. Decision rule. We evaluate H_0 directly because we can calculate the probability of chance events, but there are no mathematics for the probability of the alternative hypothesis.

    a. Chance. We calculate the probability of the obtained results if chance alone were operating.

    b. Alpha level (α). If the resulting probability turns out to be less than or equal to a critical probability level called the alpha level, we reject the null hypothesis. Common levels for the alpha are $\alpha = .05$ and $\alpha = .01$. When we reject H_0, the results are said to be significant or reliable.

II. Decision Errors

A. <u>Type I error.</u> This is when the null hypothesis is rejected when in fact the null hypothesis is true.

B. <u>Type II error.</u> This is when one retains the null hypothesis when in fact the null hypothesis is false.

C. <u>Trade-offs.</u> The probability of making a Type I error is set by alpha. The probability of making a Type II error is called beta. If we make alpha more stringent, we increase beta. Setting alpha and beta depends upon what it would cost to make either a Type I or Type II error.

III. Evaluating the Tail of the Distribution

A. <u>Directional hypothesis.</u> If the alternative hypothesis is directional, we determine the probability of getting the obtained outcome or any even more extreme in the direction hypothesized. We evaluate the tail of the distribution.

B. <u>Nondirectional hypothesis.</u> If H_1 is nondirectional, we evaluate the obtained result or any even more extreme in both directions (both tails).

C. <u>Two-tailed probability evaluations.</u> If the experimenter has no valid basis for directional hypothesis, one uses a two-tailed evaluation.

D. <u>One-tailed probability evaluation.</u> If the experimenter has a good theoretical basis and data from other studies, a directional hypothesis may be used and the experimenter may use a one-tailed probability evaluation.

IV. Summary of the Sign Test

A. Used only in replicated measures design.

   1. Null and alternative hypotheses generated.

   2. Directional or nondirectional.

   3. Alpha level set.

B. Difference between control and experimental conditions is calculated.

   1. Sign of the difference is recorded.

   2. Magnitude of difference is ignored

   3. Ties are ignored.

   4. Assumptions of binomial distributions must be satisfied.

C. Calculate the probability of getting the obtained results and results more extreme using one or two tails of the binomial distribution depending on the nature of alternative hypothesis.

D. Compare resulting probability to alpha.

E. Draw conclusions and generalizations (with the appropriate cautions).

CONCEPT REVIEW

When a scientist has an idea, he can generate a (1) _____

(1) hypothesis

which can be tested in an (2) _____. An experiment allows the

(2) experiment

scientist to test the influence of the (3) _____ variable upon

(3) independent

the (4) _____ variable while controlling for the (5) _____ of

(4) dependent
(5) influence

other variables.

    When a scientist asks an experimental question it generally

centers around an (6) _____ he has developed. The

(6) hypothesis

(7) _____ hypothesis states that the independent variable

(7) alternative

(8) _____ the dependent variable. In every experiment, besides

(8) affects

the alternative hypothesis, there exists the (9) _____ hypothesis.

(9) null

The null hypothesis for a nondirectional alternative hypothesis

asserts that the independent variable has (10) _____ effect

(10) no

upon the dependent variable and the observed results are due

to (11) _____. These two hypothesis are logical counterparts to

(11) chance

each other. They are (12) _____ (13) _____ and

(12) mutually
(13) exclusive

(14) _____. If one hypothesis is true the other must be

(14) exhaustive

(15) _____. If one is false the other must be (16) _____.

(15) false
(16) true

The alternative hypothesis is symbolized by (17) _____.

(17) H_1

The null hypothesis is symbolized by (18) _____. The

(18) H_0

alternative hypothesis can be either (19) _____ or

(19) directional

(20) _____. In a directional hypothesis the experimenter

(20) nondirectional

specifies the (21) _____ of the effect of the independent

(21) direction

variable upon the dependent variable. In a nondirectional

hypothesis the experimenter merely states that there is an

(22) _____, but not in which (23) _____.

(22) effect
(23) direction

    After the experimenter conducts an experiment, he evaluates

the (24) _____ hypothesis by determining the (25) _____

(24) null
(25) probability

of obtaining the observed results or results even more extreme if chance alone is at work. The reason we evaluate the null hypothesis is because we can calculate the probability of (26) _____   (26) chance
events but there are no mathematics worked out for the

27) _____ hypothesis. However, if we can show that the   (27) alternative

null hypothesis is (28) _____, then the alternative must be true.   (28) false

To reject the null hypothesis, one must make a decision depending on the likelihood of the observed results (tail evaluation). If the probability of observing the outcome of the study is too low to be reasonably explained by chance, we can reject H_0.

By convention, if the probability of observing the results by

chance is equal to or less than (29) _____ or (30) _____,   (29) .05
(30) .01

we reject (31) _____. The value .05 or .01 is called the   (31) H_0

(32) _____ (33) _____. If the probability of observing the   (32) alpha
(33) level

obtained results is (34) _____ or (35) _____ the (36) _____   (34) equal to
(35) less than
(36) alpha

level, we reject the (37) _____ hypothesis. If the alpha level   (37) null

equals .05 then the null hypothesis would be rejected if the

observed results occur with a probability equal to or less than

(38) _____ times out of (39) _____ assuming (40) _____   (38) five
(39) 100
alone.   (40) chance

When one evaluates the outcome of an experiment one tries to determine the state of reality concerning the independent variable. An experiment allows us to make an inference about that reality but our inferences could be wrong. If we reject H_0 when H_0 is false we have made a (41) _____ decision. If we retain H_0 (41) correct

when H_0 is true, we have also made a (42) _____ decision. If   (42) correct

we reject H_0 when H_0 is true, we have made a (43) _____ error.   (43) Type I

If we retain H_0 when H_0 is false, we have made a (44) _____   (44) Type II

error. The probability of a Type I error is determined by

(45) _____. The probability of a type II error is given by   (45) alpha

(46 _____. Unfortunately, changing alpha affects   (46) beta

(47) _____. As alpha is made more stringent, beta   (47) beta

(48) _____. Thus, decreasing the probability of a Type I error   (48) increases

increases the probability of a Type (49) _____ error. (Life is like   (49) II

that sometimes.)

   In analyzing the results of a particular study, it is (50) _____   (50) incorrect

to analyze just the specific outcome. Instead, we must determine

the (51) _____ of getting the specific outcome or any even   (51) probability

more (52) _____. It is this probability we compare to   (52) extreme

(53) _____ to assess the reasonableness of the (54) _____   (53) alpha
                                                                (54) null
(55) _____. This is to say that we evaluate the (56) _____   (55) hypothesis
                                                                (56) tail
of the distribution. If H_1 is nondirectional we evaluate

(57) _____ tails.   (57) both

   As an example of a statistical inference test we have

introduced the (58) _____ test. It is used in (59) _____   (58) sign
                                                                (59) repeated
(60) _____ designs. Data are generally collected under   (60) measures

a (61) _____ and (62) _____ condition most often on the   (61) control
                                                                (62) experimental
same (63) _____. The (64) _____ between these con-   (63) subjects
                                                                (64) difference

ditions is calculated and the number of (65) _____ and

(66) _____ are recorded. The (67) _____ of the differences

is ignored. The sign test is not very (68) _____ because it

does ignore the magnitude of the differences. In the case of ties,

the ties are (69) _____. One then evaluates the number of

pluses (or minuses) just like heads (or tails) using the

(70) _____ distribution. This presupposes all the (71) _____

of the binomial distribution. One then calculates the probability

of getting the obtained results or results (72) _____

(73) _____. Whether one evaluates one or both

(74) _____ of the distribution depends on whether the alterna-

tive hypothesis is (75) _____ or (76) _____. One then

compares the resulting probability with (77) _____. If the

obtained probability is less than (78) _____ we (79) _____

(80) _____. If one rejects the null hypothesis, one is at risk

of making a (81) _____ error. If one retains the null hypothesis

one may be making a (82) _____ error. Notice that we always

reject or retain the (83) _____ hypothesis. Finally, one can

draw conclusions and offer generalizations. One can only

generalize to the (84) _____ from which the (85) _____ was

drawn.

(65) pluses
(66) minuses
(67) magnitude
(68) sensitive

(69) discarded

(70) binomial
(71) assumptions

(72) more

(73) extreme

(74) tails

(75) directional
(76) nondirectional
(77) alpha

(78) alpha
(79) reject
(80) H_0

(81) Type I

(82) Type II

(83) null

(84) population
(85) sample

EXERCISES

1. A teacher wanted to compare two methods for teaching math. He randomly divided the class in half. Half the class was given the standard method first, followed by the new method. The other half was given the two methods in the opposite order. The results of tests given after each of the two methods are summarized below:

Student	Standard Method	New Method
1	80	82
2	78	79
3	65	75
4	92	100
5	85	85
6	60	61
7	64	60
8	82	80
9	90	92
10	78	86
11	70	79
12	85	90
13	86	80

a. State the nondirectional alternative hypothesis.
b. State the null hypothesis.
c. Using $\alpha = .05_{2 \text{ tail}}$, what do you conclude?
d. What type of error might you be making?

2. A panel of consumers was asked to rate "Natural Vitaglo" breakfast cereal before and after a change in manufacturing techniques. A higher number reflects a higher preference for the cereal. The results are given below:

Rater	Old Process	New Process
A	6	10
B	7	9
C	7	8
D	8	10
E	5	6
F	9	10
G	10	9
H	4	5
I	6	8
J	3	4

    a. State the nondirectional alternative hypothesis.
    b. State the null hypothesis.
    c. What do you conclude using $\alpha = .05_{2\ tail}$?
    d. What type of error might you be making?
    e. What would your conclusion be if $\alpha = .01_{2\ tail}$ had been specified?
    f. Why did the investigator use a 2-tailed test?

3. Why do we test the null hypothesis instead of the alternative hypothesis?

4. A young man in a pinstriped suit, black shirt and white tie comes up to you and asks if you want to flip a coin for a dollar. Feeling lucky you agree. You call "tails" eight times in a row and you lose each time. You begin to wonder if the coin is biased.

    a. State the directional alternative hypothesis.
    b. State the null hypothesis.
    c. What do you conclude using $\alpha = .05_{1\ tail}$?
    d. Why did you start playing with that guy anyway?

5. If you were a scientist who believes she has discovered an inexpensive new energy source that doesn't pollute, and you designed an experiment to assess whether or not this energy source actually worked, would you rather protect yourself from making a Type I or Type II error? Why?

6. A social scientist wants to determine if showing a film will affect religious tolerance. A sample of freshman males was drawn from a university. Pre and post film ratings of religious tolerance were obtained. A higher number represents higher tolerance. The following data were obtained:

Person	Pre Film	Post Film
1	50	89
2	52	92
3	61	93
4	70	68
5	58	95

a. State the nondirectional alternative hypothesis.
b. State the null hypothesis.
c. What do you conclude using $\alpha = .01_{2\ tail}$?
d. To what population can you generalize?
e Are there any results from a sample of the size which would have allowed you to reject H_0?

7. A pharmacologist recently developed a drug which, in theory, should have an effect on the activity of the lateral hypothalamus. It is likely that such a drug would have an effect on appetite. A small pilot experiment is designed to test this hypothesis.

   a. State H_0 and H_1.
   b. At what level would you set alpha? Why?
   c. Would you use a one- or two-tailed test? Why?

8. An after-shave company recently advertised that if you use their product you will have more sex appeal. To test this claim you enlist the assistance of 12 men from a local social club. The men are either given the new after-shave or colored water. After one month the type of liquid each man had been using was changed to the other type; i.e., water changed for after-shave or after-shave for water. The experiment continued for another month. The dependent variable was the number of dates each man had for each month. The results are shown below:

Subject	1	2	3	4	5	6	7	8	9	10	11	12
Colored water	7	9	4	2	8	3	6	5	10	1	0	4
After-Shave	6	5	10	3	9	7	5	5	11	2	1	5

a. State H_0 and H_1.
b. What do you conclude from the data shown above? Use $\alpha = .05_{2\ tail}$
c. What type of error might you have made?
d. What is one possible explanation for these results?

9. The following probabilities are the results of several different experiments.

Experiment	p(obtained)$_{2\ tail}$
1	.01
2	.04
3	.05
4	.17
5	.20
6	.006

   a. What would you conclude for each experiment if $\alpha = .05_{2\ tail}$?

   b. What would you conclude if $\alpha = .01_{2\ tail}$?

10. What are the possible costs to making a Type I error in a study that tests the null hypothesis that a new type of male contraceptive pill is not effective? (The alternative hypothesis is that it is effective.)

Answers: 1a. The new teaching method affects test scores.   **1b.** The new teaching method has no effect on test scores. $P = Q = .50$.   **1c.** Retain H_0 since .1458 is greater than .05. This experiment fails to establish that there is any difference in effectiveness between the new and standard teaching methods.   **1d.** Type II error. (NOTE: did you remember to omit case 5 because of tied scores?)

2a. The new manufacturing process affects the taste of "Natural Vitaglo" as assessed by preference ratings.   **2b.** The new manufacturing process has no effect on the taste of "Natural Vitaglo" as assessed by preference ratings. $P = Q = .50$.   **2c.** Since .0216 < .05, reject H_0 in favor of H_1. The change in process affects taste.   **2d.** Type I error. **2e.** Retain H_0.   **2f.** Nondirectional H_1.

3. Because the mathematics have only been established for chance events.

4a. The coin is biased in favor of producing heads.   **4b.** The coin is not biased in favor of producing heads. $P \leq Q \leq .50$.   **4c.** Reject H_0 since .0039 is less than .05   **4d.** Who knows?

5. There is no correct answer but probably you would wish to protect yourself against a Type II error since if H_0 is false you definitely want to reject it and it seems there are more costs to making a Type II error than a Type I error. On the other hand, to avoid misleading the public, you would probably want to set alpha to at least .05.

6a. The film affects religious tolerance.   **6b.** The film has no effect on religious tolerance and the results are due to chance alone.   **6c.** Retain H_0 since .3718 > .01.

The experiment fails to show that the film affects religious tolerance.   **6d.** Freshman men at that university   **6e.** No, it's a very poor experiment.

7a. H_0: the new drug has no effect on food intake. Any effect is due to chance alone. H_1: taking this new drug effects food intake. The differences are such that they cannot reasonably be attributed to chance.   **7b.** $\alpha = .10$. During a pilot study one generally wishes to set alpha at a less stringent level than perhaps under other circumstances to see if a larger scale trial is warranted. There is no correct answer to this question. You should have a reasonable rationale for whatever alpha level you choose.   **7c.** Two-tailed. From the information given there is no justification for selecting a directional hypothesis.

8a. H_1: using the after-shave has an effect on the number of dates men who wear it obtain. H_0: the use of the after-shave has no effect on the number of dates men who wear it obtain. $P = Q = .50$.   **8b.** $p = .2268$ which is $> \alpha$; therefore retain H_0.   **8c.** Type II   **8d.** The number of dates might not be a good measure of sex appeal. There are a lot of other possible answers.

9a. Reject H_0 for 1, 2, 3, and 6; retain H_0 for 4 and 5.   **9b.** Reject H_0 for 1 and 6; retain for 2, 3, 4, and 5.

10. Increase in birth rate, possible increase in marriage rate, unplanned pregnancies, etc., etc.

TRUE-FALSE QUESTIONS

T F   1. The sign test ignores the magnitude of the difference scores.

T F   2. The sign test analyzes raw scores.

T F   3. It impossible to get 20 pluses out of 20 pairs of scores in a replicated measures design due to chance alone.

T F   4. A replicated measures design is the same thing as a correlated groups design.

T F   5. Unless the very same subject is used in the control and experimental condition the design cannot be a replicated measures design.

T F   6. The statement "Drug X has no effect on Y and any observed effect is due to chance alone" is an example of a nondirectional alternative hypothesis.

T F   7. H_0 and H_1 must be mutually exclusive and exhaustive.

T F   8. H_0 always asserts that the dependent variable has no effect on the independent variable.

T F 9. It is not possible to analyze the probability of the alternative hypothesis because the laws of probability are derived for chance events. This is why we test H_0 instead of H_1.

T F 10. If we reject H_0 one can say the experimental results are significant.

T F 11. If we reject H_0 then we are absolutely certain the independent variable affected the dependent variable.

T F 12. The critical probability for rejecting H_0 is the alpha level.

T F 13. Retaining H_0 and accepting H_0 mean the same thing.

T F 14. Type I errors are always worse to make than Type II errors.

T F 15. If H_0 is true and you retain H_0 in your experiment you have made a Type II error.

T F 16. If the probability of the results obtained equals .05 and $\alpha = .05$, you should reject H_0.

T F 17. By making alpha smaller we can decrease the probability of making a Type I error.

T F 18. Beta is the probability of retaining H_0 when H_0 is false.

T F 19. The more stringent the alpha level is, the easier it is to detect an effect of the independent variable on the dependent variable.

T F 20. If $\alpha = .05$ and chance alone is operating, it is reasonable to expect that on the average one would make one Type I error in every 20 experiments.

T F 21. As the probability of making a Type I error goes down by making α more stringent, the probability of making a Type II error goes up.

T F 22. In the real world the scientist can never be certain whether or not H_0 is true.

T F 23. The probability of making a Type I error increases with independent replication.

T F 24. In hypothesis testing it is incorrect to evaluate the probability of the specific outcome of the experiment.

T F 25. If H_1 is nondirectional one must evaluate the probability of getting the obtained result or a result more extreme in both directions to properly test H_0.

T F 26. It is appropriate to determine if H_1 is directional or nondirectional after the data is analyzed.

T F   27. If $\alpha = .01$ and H_1 is 2-tailed then there is .005 under each tail.

Answers: 1. T   **2.** F   **3.** F   **4.** T   **5.** F   **6.** F   **7.** T   **8.** F   **9.** T   **10.** T
11. F   **12.** T   **13.** F   **14.** F   **15.** F   **16.** T   **17.** T   **18.** T   **19.** F   **20.** T
21. T   **22.** T   **23.** F   **24.** T   **25.** T   **26.** F   **27.** T.

SELF-QUIZ

1. If p (obtained) from an experiment equals .05 and alpha equals .05 (both two-tailed), what would you conclude?

    a. reject H_0
    b. retain H_0
    c. reject H_1
    d. retain H_1

2. If you reject the null hypothesis, what type of error might you be making?

    a. Type I
    b. Type II
    c. Type III
    d. cannot be determined

3. If alpha equals .05, how many times out of 100 would you expect to reject the null hypothesis when the null hypothesis is in fact true?

    a. 1
    b. 0.05
    c. 0.01
    d. 5

4. If we drew a random sample from an introductory psychology class, to whom could we generalize our results?

    a. all of human kind
    b. the university
    c. all psychology students
    d. the students in that introductory psychology class

5. If we set alpha at .05 instead of .01, other factors held constant _____.

    a. we have a greater risk of a Type I error
    b. we have a greater risk of a Type II error
    c. we have a lower risk of a Type II error
    d. a and c

6. If you reject H_0 when H_0 is false, you have made a _____.

    a. Type I error
    b. Type II error
    c. correct decision
    d. none of the above

7. If the alpha level is changed from .05 to .01, what effect does this have on beta?

    a. beta decreases
    b. beta increases
    c. beta is unaffected
    d. cannot be determined

8. The sign test can be used for _____.

    a. a repeated measures design
    b. a replicated measures design
    c. a correlated measures design
    d. all of the above

9. One can say it is always preferable to make a Type II error.

    a. True
    b. False
    c. It depends on the costs of making a Type I or Type II error

10. In a nondirectional alternative hypothesis, evaluating the probability of observing 7 pluses out of 8 events equals _____.

    a. $p(7)$
    b. $p(7) + p(8)$
    c. $p(0) + p(1) + p(7) + p(8)$
    d. $p(0) + p(8)$

Consider the following hypothetical data collected using replicated measures design:

Subject	1	2	3	4	5	6	7	8	9	10
Pre	50	49	37	16	80	42	40	58	31	21
Post	56	50	30	25	90	44	60	71	32	22

11. In a two-tailed test of H_0 using $\alpha = .05$, what is p(obtained) for the results shown?

   a. .0500
   b. .0108
   c. .1094
   d. .0216

12. What would you conclude using the information in problem 11?

   a. reject H_0
   b. accept H_0
   c. retain H_0
   d. retain H_1
   e. b or c

13. What would you conclude using $\alpha = .01_{2\ tail}$?

   a. reject H_0
   b. accept H_0
   c. retain H_0
   d. fail to reject H_1
   e. b or c

14. What type error might you be making using $\alpha = .05_{2\ tail}$?

   a. Type I
   b. Type II
   c. Type III
   d. cannot be determined

Answers: 1. a   **2.** a   **3.** d   **4.** d   **5.** d   **6.** c   **7.** b   **8.** d   **9.** c   **10.** c   **11.** d   **12.** a   **13.** c   **14.** a.

11 | POWER

CHAPTER OUTLINE

I. Definitions

  A. Power is the probability that the results of experiment will allow rejection of the null hypothesis if the independent variable has a real effect.

  B. Power is the probability that the results of experiment will allow rejection of the null hypothesis if the null hypothesis is false.

  C. Power is the probability of making a correct decision when H_0 is false.

  D. P_{real} is the probability of a plus with any subject in the sample of the experiment when the independent variable has a real effect. It is also the proportion of pluses in the population if the experiment were done on the entire population and the independent variable has a real effect..

  E. P_{real} varies with the magnitude and direction of the real effect.

  F. P_{null} is the probability of getting a plus with any subject in the sample of the experiment when the independent variable has no effect.

  G. $P_{null} = .50$ (H_1 nondirectional).

H. P_{real} equals any value other than .50 (H_1 nondirectional)

II. Power and Beta (β)

A. Power = 1 - beta.

B. Beta = 1 - power.

C. As the power of an experiment increases, the probability of making a Type II error decreases.

D. Maximizing power minimizes beta.

E. Methods of increasing power.

1. Increase the magnitude of effect of independent variable.

2. Increase sample size.

III. Identifying Correct State of Reality

A. Set alpha stringently.

B. Set beta low (maximize power).

IV. Calculation of Power is a Two-Step Process

A. Assume the null hypothesis is true (P_{null} = .50) and determine the possible sample outcomes in the experiment which allow H_0 to be rejected.

B. For the value of P_{real} under consideration (e.g. P_{real} = .40) determine the probability of getting any one of the above sample outcomes. This probability is the power of the experiment to detect the specific magnitude of effect.

CONCEPT REVIEW

(1) _____ is the probability of making a Type II error. By maximizing (2) _____ we can minimize beta. The power of an experiment is a measure of the (3) _____ of the experiment to detect the real (4) _____ of the (5) _____ variable. A more

(1) Beta

(2) power

(3) sensitivity

(4) effect
(5) independent

formal definition of power is the (6) _____ of rejecting

(7) _____ when the (8) _____ variable has a (9) _____.

effect. P_{real} is the probability of getting a plus if the independent

variable has a (10) _____ effect. P_{null} is the probability of getting

a plus if the independent variable has (11) _____ effect.

P_{real} is a measure of the (12) _____ and (13) _____ of

of the real effect of the (14) _____ variable.

The further (15) _____ is from .50, the (16) _____ is

the effect of the independent variable. The closer the value

of (17) _____ is to .50, the (18) _____ is the

(19) _____ effect of the independent variable. If P_{real}

= .3 the effect is (20) _____ than for P_{real} = .2. In a

given experiment, the greater the (21) _____ of the independent

variable the greater the (22) _____.

To calculate power we first assume that H_0 is (23) _____

(P_{null} = (24) _____). Then determine all the (25) _____

outcomes in the experiment which allow H_0 to be (26) _____

Second, for the magnitude of real effect under consideration,

we determine the (27) _____ of getting any one of the above

sample outcomes. This probability is the (28) _____ of the

experiment to detect this level of (29) _____ effect. For each

magnitude of real effect, we must calculate a new value for

(30) _____. If the power of an experiment is .6500, that

means that we have a (31) _____ % chance of rejecting

(6) probability

(7) H_0
(8) independent
(9) real

(10) real

(11) no

(12) magnitude
(13) direction
(14) independent

(15) P_{real}
(16) larger

(17) P_{real}
(18) smaller
(19) real

(20) smaller

(21) effect

(22) power

(23) true

(24) .50
(25) sample
(26) rejected

(27) probability

(28) power

(29) real

(30) power

(31) 65

H_0 when H_0 is false and a (32) _____ % chance of

making a Type II error. If H_0 is true the probability of re-

jecting the null hypothesis is determined by (33) _____.

    In an experiment there are only two possibilities. Either H_0

is (34) _____ or it is (35) _____. By minimizing (36) _____

and (37) _____ we maximize the likelihood our conclusions will

be (38) _____. Power = 1-(39) _____. One way to achieve

a low beta when alpha is set at a stringent level is to have a larger

(40) _____.

    If one obtains nonsignificant results in an experiment, one

(41) _____ to reject H_0 . This may be because H_0 is

(42) _____ or because H_0 is false, but the (43) _____of the

experiment was (44) _____. Therefore, we (45) _____.

accept H_0 Rather, we (46) _____ it as a reasonable possible

explanation of the results. A powerful experiment will yield a high-

er likelihood of rejecting H_0 only if H_0 is (47) _____.

    A power analysis is useful when initially (48) _____ an

experiment or when interpreting the (49) _____ results from

an experiment. Since we never know the magnitude of the effect

of the independent variable before carrying out the experiment, we

usually (50) _____ a value for P_{real} based on pilot work or

similar studies.

Answer key	
(32)	35
(33)	alpha
(34)	true
(35)	false
(36)	alpha
(37)	beta
(38)	correct
(39)	beta
(40)	N (sample size)
(41)	fails
(42)	true
(43)	power
(44)	low
(45)	do not
(46)	retain
(47)	false
(48)	designing
(49)	nonsignificant
(50)	estimate

EXERCISES

1. A professor thinks that a certain movie is fun and exciting for students. He selects heart rate (HR) measured before and after the movie as the dependent variable for a study on 12 students. He figures if the students are having fun and are excited, the heart rate will be higher right after the movie. The following data have been obtained.

Student		1	2	3	4	5	6	7	8	9	10	11	12
Pre-movie HR		70	68	74	88	75	79	95	65	87	73	69	96
Post-movie HR		72	69	71	92	74	85	100	80	90	77	71	88

   a. Based on these results, what should the professor conclude using $\alpha = .05_{2\ tail}$? (Assume that heart rate is a good measure of fun and excitement.).
   b. If the effect of the independent variable were such that $P_{real} = .60$, what is the power in this experiment?
   c. What type error might the investigator be making?

2. In an experiment where the experimental effect is such that $P_{real} = .70$ and N = 14, what is the power of the experiment? Use $\alpha = .05_{2\ tail}$ with the sign test.

3 Assume you have done an experiment and used the sign test to analyze the results. N = 14.

   a. If your experimental treatment was moderately effective ($P_{real} = .70$), what was the power of your experiment? Assume $\alpha = .05_{1\ tail}$.
   b. Generalize from questions 2 and 3a about alpha and power.

4 Assume you have N = 15, $\alpha = .05_{1\ tail}$, and applied the sign test to the data.

   a. What is the power if $P_{real} = .80$?
   b. If you increased the sample size to 20, what would be the power?
   c. What is the value of beta if N = 20?
   d. Generalize from questions 4a and 4b about N and power.

5. You are considering testing a new drug which is supposed to facilitate learning in mentally retarded children. Based on preliminary research, you have some idea about the magnitude of the drug's effect. Because of your work schedule you can either do the test with 15 subjects or with 20. You would like to only run 15 subjects, but if running 20 will make power at least 20% higher, you will run 20 subjects. Calculate power for the following conditions:

   a. N = 15, $\alpha = .05_{1\ tail}$, $P_{real} = .70$.

b. $N = 20$, $\alpha = .05_{1\text{ tail}}$, $P_{real} = .70$.

c. Based on your calculations, how many subjects will you test?

6. A wine company has retained you as a consultant to help evaluate the taste preference for their wine compared to the market leader. You plan to give 20 subjects a taste of both wines and ask them to rate the wines on a 10 point scale. The order of presentation is randomized so there is no order effect. Your plan is to analyze the results using the sign test. If the magnitude of the difference is such that $P_{real} = .80$, what is the power of your experiment to detect a preference for one wine or the other? Use $\alpha = .05_{2\text{ tail}}$.

7. Ajax Pest Control Company has bred a new strain of pest-eating insects for the garden. In several test gardens the number of pests before and after introduction of their new insect has been recorded. The following data were obtained:

Garden	1	2	3	4	5	6	7	8
Pre	32	67	83	22	39	44	90	11
Post	26	12	34	66	10	0	17	25

a. What is the power of this experiment using $\alpha = .01_{2\text{ tail}}$ if $P_{real} = .70$?

b. How could you increase the power?

8. In question 7, what is the probability of making a Type II error?

9. In an experiment, if the analysis of the data does not allow us to reject the null hypothesis, this can only be so because we do not have enough experimental power to reject H_0. True or false? Why?

10. In an experiment the calculated power to detect an effect is .4600. What is the value for beta?

11. In an experiment the value for beta has been shown to be .7503. What is the power of this experiment?

Answers: 1a. Retain H_0 because .1458 > .05 **1b.** .0863 **1c.** Type II **2.** .1609 **3a.** .3552 **3b.** When the effect is in the predicted direction, a one-tailed alpha level results in higher power than a two-tailed alpha level. **4a.** .6481 **4b.** .8042 **4c.** .1958 **4d.** As N increases, power increases **5a.** .2968 **5b.** .4163 **5c.** 20 subjects **6.** .8042 **7a.** .0577 **7b.** Increase N; increase magnitude of effect. **8.** .9423 **9.** false, it is also possible H_0 is true **10.** .5400 **11.** .2497.

TRUE-FALSE QUESTIONS

T F   1. If H_0 is true, power is the probability of not making a Type II error.

T F   2. If in reality H_0 is true, then designing a very powerful experiment (without changing alpha) will allow us to reject H_0 more readily.

T F   3. Power is the probability of making a correct decision when H_0 is true.

T F   4. If the independent variable is more powerful in experiment A than in experiment B, there is a higher probability of rejecting H_0 in A than in B if, in fact, H_0 is false.

T F   5. H_0 states that the independent variable has no effect. H_1 is nondirectional.

T F   6. It is impossible for an experiment to have power equal to zero.

T F   7. An experiment where $P_{real} = .10$ is less likely to allow rejection of H_0 if H_0 is false than one where $P_{real} = .80$.

T F   8. The greater the sample size the greater the power.

T F   9. Power = $1 - \beta$

T F   10. Beta is the probability of retaining H_0 when H_0 is false.

T F   11. $\beta = 1 - \alpha$.

T F   12. A good experiment is one where both α and β are low.

T F   13. If one conducts an experiment and ends up retaining H_0, it is probably because the experiment was not powerful enough.

T F   14. In an experiment where we retain H_0, one cannot be certain if it was because H_0 was true or that the experiment was not powerful enough to detect an effect by the independent variable.

T F   15. The concept of statistical power applies only to the sign test.

T F   16. An experiment with $\alpha = .05$ is more powerful than one where $\alpha = .01$, other factors held constant.

T F   17. Increasing N, making α more stringent, and increasing the effect of the independent variable all increase power.

SELF-QUIZ

1. If the independent variable has a real effect, the probability of rejecting H_0 is _____.

   a. power
   b. 1 - power
   c. alpha
   d. beta

2. One can increase power by _____.

   a. increasing N
   b. making alpha more stringent
   c. increasing the effect of independent variable
   d. a and c

3. If the power of an experiment is .3400, the probability of retaining H_0 when H_0 is false is _____.

   a.   .6600
   b.   .3400
   c.   .5000
   d. 1.0000

4. Which of the following P values represents the strongest effect?

   a. $P_{null} = .50$
   b. $P_{real} = .70$
   c. $P_{real} = .80$
   d. $P_{real} = .10$

5. If beta = .7500, what is the power of the experiment?

   a.   .7500
   b.   .5000
   c.   .2500
   d. 1.0000

6. Which of the following represents the null hypothesis condition for a nondirectional H_1?

   a. $P_{real} = .49$
   b. $P_{null} = .50$
   c. $P_{real} = .51$
   d. all of the above

7. In the case of $N = 3$, $P_{real} = .90$, $\alpha = .05_{1\ tail}$, using the sign test, the power is _____.

   a. 1.0000
   b. .1250
   c. .8750
   d. .0000

8. Which of the following may be <u>false</u>?

   a. alpha + beta = 1
   b. power + beta = 1
   c. 1 - power = beta
   d. 1 - beta = power

9. In a two-tailed test, which of the following would yield the same power as a magnitude of effect represented by $P_{real} = .90$?

   a. $P_{real} = 1.00$
   b. $P_{real} = .10$
   c. $P_{real} = .80$
   d. none of the above

10. If H_0 is false the probability of making a correct decision is _____.

    a. alpha
    b. 1 - alpha
    c. power
    d. 1 - power

11. Assume that you have conducted an experiment and the tail probability for the results you obtained was .0900 and $\alpha = .05$. You can conclude _____.

    a. the independent variable has no effect
    b. the independent variable has a weak effect
    c. you were unable to detect an effect of the independent variable with your experiment
    d. the null hypothesis is false

12. Truth demands that scientists set _____.

    a. alpha equal to .05
    b. alpha equal to .01
    c. a or b
    d. not necessarily $\alpha = .01$ or .05, but use their best judgment

13. In the sign test, if H_0 is false, then P_{real} _____.

    a. equals .05
    b. equals .50
    c. equals 1 - .05
    d. does not equal .50

14. In an experiment, if the effect of the independent variable and alpha remain the same and N increases, _____.

    a. power increases
    b. power decreases
    c. power remains the same
    d. cannot be determined

15. In an experiment with N = 14 and $P_{real} = .10$, what is the power using $\alpha = .05_{2\ tail}$?

    a. .8417
    b. .0132
    c. .0000
    d. .0500

16. In the same experiment as question 15, what is the power using $\alpha = .01_{2\ tail}$?

    a. .0500
    b. 1.0000
    c. .5847
    d. .9868

17. An experiment with N = 18 is more powerful than an experiment with N = 17, all other things being the same.

    a. true
    b. false

18. In order to calculate power you must know _____.

    a. alpha
    b. N
    c. P
    d. the results of the experiment
    e. all of the above
    f. a, b and c

19. In an experiment with N = 6 and P_{real} = .15, what is the power using $\alpha = .01_{2\,tail}$?

    a. 1.0000
    b. .0500
    c. .0000
    d. .3771

Answers: 1. a **2.** d **3.** a **4.** d **5.** c **6.** b **7.** d **8.** a **9.** b **10.** c
11. c **12.** d **13.** d **14.** a **15.** a **16.** c **17.** a **18.** f **19.** c.

12 | Mann-Whitney U Test

CHAPTER OUTLINE

I. **Independent Groups Design**

A. Characteristics.

1. Each subject tested only once.

2. Subjects are randomly sampled from the population and then randomly divided into 2 or more groups.

3. All subjects in one group run under experimental conditions.

4. All subjects in other group run under control conditions.

B. Data analysis.

1. No basis for pairing scores between conditions.

2. Differences between groups are compared to determine if chance alone is reasonable explanation for the differences between group scores.

3. Analyze the two groups of scores as separate samples.

II. Analysis Using the Mann-Whitney U Test

A. This test analyzes the separation between the two sets of sample scores and allows us to determine the probability of getting the obtained separation or even greater separation if chance alone is at work.

B. The greater the separation between two sets of scores the less likely they are both random samples drawn from the same population.

C. The greater the overlap between two sets of scores, the more reasonable chance alone explains the observed separation and samples are random samples from same population.

III. Calculation of Separation (U or U')

A. Count the total number of C (control) scores that are lower than E (experimental) scores, or count the number of E scores lower than C scores when scores are listed in rank order. This method cannot be used where ties exist.

B. Equations for U and U'.

$$U_{obt} = n_1 n_2 + \frac{n_1 (n_1 + 1)}{2} - R_1$$

$$U_{obt} = n_1 n_2 + \frac{n_2 (n_2 + 1)}{2} - R_2$$

where U_{obt} = a statistic which indicates the degree of separation between the two samples.
n_1 = number of scores in group 1
n_2 = number of scores in group 2
R_1 = the sum of ranks for scores in group 1
R_2 = the sum of ranks for scores in group 2

Note: One formula will yield U and the other U'. U is the smaller and U' the larger of two numbers indicating the degree of separation between the two samples. Calculate both numbers and the smaller is U and the larger is U'. The equations can be used anytime. Tied scores are treated the same as tied scores in the Spearman rho.

C. Values of U.

1. A value of zero represents greatest degree of separation possible. No overlap exists.

2. Larger values of U indicate greater overlap. When U = U', the greatest overlap exists.

3. **$U_{obt} + U'_{obt} = n_1 n_2$.** Knowing this means one only has to calculate either U or U' and the other value can be readily determined assuming the calculations are correct.

IV. Determining the Probability of U if Chance Alone is Operating

A. Probability distribution for U.

1. All probability distributions are symmetrical regardless of n_1 and n_2.

2. As n_1 and n_2 increase the distribution approaches normality.

3. Probabilities can be obtained by developing a probability distribution for each value of n_1 and n_2 or by using tables C_1-C_4.

B. Tables for evaluating U and U'.

1. Find appropriate table for alpha.

2. Upper entry is highest value of U for various n_1 and n_2 combinations that will allow rejection of H_0.

3. Lower entry is lowest value of U' that will allow rejection of H_0.

4. Locate U value at the intersection of n_1 and n_2.

5. Draw conclusions.

V. Practical Considerations for Using the Mann-Whitney U Test

A. Use does not depend on shape of population of scores.

B. Data must be at least ordinal.

C. Can be substituted for Student's t test when assumptions of test are not met.

D. Not as powerful as Student's t test but still quite powerful.

E. Used for independent groups design.

CONCEPT REVIEW

There are essentially two basic experimental designs used in studying behavior. The first we studied was the (1) _____ (2) _____ design. In that design the essential feature is that there are (3) _____ scores and the (4) _____ between scores are analyzed to see if (5) _____ alone can reasonably explain them.

The other type is called the (6) _____ (7) _____ design. Here each subject is tested only (8) _____. This design has (9) _____ or more groups, usually an (10) _____ group and a (11) _____ group. Subjects are (12) _____ selected from the (13) _____ and then randomly divided into the two (or sometimes more) groups. There is (14) _____ basis for pairing scores between the two conditions. A comparison is made between the scores of each (15) _____ to determine if chance alone is a reasonable explanation of the (16) _____ between the group scores. The scores are analyzed as separate (17) _____. If chance alone is the correct explanation of the results then sets of sample scores (groups) can be considered random samples from the (18) _____ populations.

The Mann-Whitney U test analyzes the (19) _____ between the two sets of (20) _____ scores and allows us to determine the (21) _____ of getting the obtained separation or even (22) _____ separation if both sets of scores are (23) _____

(1) repeated
(2) measures
(3) paired
(4) differences
(5) chance

(6) independent
(7) groups
(8) once

(9) two
(10) experimental
(11) control
(12) randomly
(13) population

(14) no

(15) group

(16) differences

(17) samples

(18) same or identical
(19) separation

(20) sample

(21) probability

(22) greater

samples from (24) _____ populations. The (25) _____

the separation between the two samples, the less likely it is that

they are random samples drawn from the same (26) _____.

The more (27) _____ between the two sample distributions

the more reasonably (28) _____ explains the results.
    The value (29) _____ is the smaller of two numbers that

indicate the degree of separation between the two samples for

the Mann-Whitney U test. (30) _____ is the larger of the two

numbers indicating the same thing. One way of calculating U or U'

is to (31) _____ the scores from both groups and (32) _____

(33) _____ them. We can then count the number of E scores

(members of the experimental group) that are (34} _____ than

C scores (members of the control group). To do this, count the

number of E scores that are lower than each C score and

(35) _____ these values. Or we could count the C's less than

E's. Whichever number was (36) _____ would be U_{obt} and the

higher number would be (37) _____. Consider the following

scores:

(23) random
(24) identical
(25) greater

(26) population

(27) overlap

(28) chance
(29) U_{obt}

(30) U'_{obt}

(31) combine
(32) rank
(33) order

(34) lower

(35) sum

(36) lower

(37) U'_{obt}

$$28 \quad 30 \quad 32 \quad 34 \quad 35 \quad 42 \quad 46$$

$$C_1 \quad C_2 \quad E_1 \quad E_2 \quad C_3 \quad E_3 \quad E_4$$

To determine U we can complete the following table:

E's less than C_1 = (38) _____	(38) 0
E's less than C_2 = (39) _____	(39) 0
E's less than C_3 = (40) _____	(40) 2
Total E's less than C's = (41) _____	(41) 2

Therefore, the value of U_{obt} is (42) _____ If we calculated the

total number of C's less than E's we would get:

C's less than E_1 = (43) _____	(43) 2
C's less than E_2 = (44) _____	(44) 2
C's less than E_3 = (45) _____	(45) 3
C's less than E_4 = (46) _____	(46) 3
Total C's less than E's = (47) _____	(47) 10

(42) 2

The value for U'_{obt} is (48) _____. This method will only work if

there are (49) _____ (50) _____ (51) _____ between

groups. As a check on our work we note that U + U' must

equal (52) _____ x (53) _____. In this case, U =

(54) _____ and U' = (55) _____. and n_1 = (56) _____

and n_2 = (57) _____. Since 12 = 12, we can be confident in

our work.

(48) 10
(49) no
(50) tied
(51) scores
(52) n_1
(53) n_2
(54) 2
(55) 10
(56) 3
(57) 4

Another, more general way to calculate U is to apply one of

the following equations:

$$\text{equation 1: } U_{obt} = n_1 n_2 + \frac{n_1(n_1+1)}{2} - R_1$$

$$\text{equation 2: } U_{obt} = n_1 n_2 + \frac{n_2(n_2+1)}{2} - R_2$$

where

n_1 = the number of scores in group = (58) _____   (58) 1

n_2 = the number of scores in group = (59) _____   (59) 2

R_1 = the (60) _____ of ranks in group 1   (60) sum

R_2 = the (61) _____ of ranks in group 2   (61) sum

If we rearranged the data in the problem above as follows, we can easily apply the equations.

1 Control Group		2 Experimental Group	
Score	Rank	Score	Rank
28	(62) _____	32	(63) _____
30	(64) _____	34	(65) _____
35	(66) _____	42	(67) _____
		46	(68) _____
R_1 =	(69) _____	R_2 =	(70) _____
n_1 =	(71) _____	n_2 =	(72) _____

(62) 1
(63) 3
(64) 2
(65) 4
(66) 5
(67) 6
(68) 7
(69) 8
(70) 20
(71) 3
(72) 4

Solving the equation for U_{obt}:

$$U_{obt} = n_1 n_2 + \frac{n_1 (n_1 + 1)}{2} - R_1$$

$$U_{obt} = (73) \_\_\_ (74) \_\_\_ + \frac{(75) \_\_\_ \times (76) \_\_\_}{2}$$

$$- (77) \_\_\_\_\_ = (78) \_\_\_\_\_$$

(73) 3
(74) 4
(75) 3
(76) 4
(77) 8
(78) 10

Since U + U' must equal n_1 (79) _____ n_2, then the U_{obt} we have just calculated must be (80) _____. The value of U must be (81) _____. This agrees with the values we obtained with

(79) times
(80) U'
(81) 2

the counting method used previously, but generally is a more useful way to calculate U and U'. If one applied equation 2 to the same data the value for U_{obt} would be (82) _____.

(82) 2

Obtaining a value for U or U' is meaningless unless we know the (83) _____ of obtaining that value or any more (84) _____ assuming chance alone is at work. This information is available in Tables C_1-C_4. The tables have cells with two entries. The upper entry is the (85) _____ value of U for various n_1 and n_2 combinations which will allow (86) _____ of H_0. The lower entry is the lowest value of (87) _____ which will allow reject-ion of H_0 at a given level of alpha.

(83) probability
(84) extreme

(85) highest

(86) rejection

(87) U'

Both U and U' measure the same degree of (88) _____ so one needs only to evaluate one or the other. If we were evaluating the value of U_{obt} using the above data at alpha = $.05_{1\ tail}$, we would use table (89) _____. For $n_1 = 3$ and $n_2 = 4$, the largest value of U that will allow rejection of H_0 is (90) _____. Since the value of U_{obt} is 2, we (91) _____ H_0. If there had been tied scores in this problem we would simply assign the tied scores a value equal to the (92) _____ of the. tied ranks.

(88) separation

(89) C_4

(90) 0
(91) retain

(92) average

The probability distribution of U for each combination of n_1 and n_2 is not difficult to understand. One simply calculates the probability of getting each possible rank ordering of (93) _____ and (94) _____ scores for the n_1 and n_2 combination. The

(93) E

(94) C

distribution of U is (95) _____ and becomes almost

(96) _____ in shape as n_1 and n_2 (97) _____. The values

of U and U' listed in Tables C_1-C_4 are taken from the appropriate

probability distribution of U.

EXERCISES

1a. What does the blank space in the entry for $n_1 = 3$ and $n_2 = 3$ in table C_3 mean?

1b. What is the power of $n_1 = 3$, $n_2 = 3$ for $\alpha = .05$ 2 tail?

2. What is (are) the only possible order(s) of C's and E's for U = 0 with $n_1 = n_2 = 4$?

3. Consider any of the tables C_1 - C_4 for U. How might one increase power in an experiment?

4. If $R_1 = 80$, $n_1 = 10$ and $n_2 = 6$, what is the value of U_{obt}?

5a. In the case where you have 3 C's and 4 E's, what is the probability of getting the order CCCEEEE?

5b. What is the value of U for this order?

6a. If U = U', what can you say about the degree of overlap of the control and experimental distributions?

6b. What would you do with H_0 in this case?

7. An animal geneticist is trying to pick an appropriate species of fish for repopulating a lake. He wants to compare how long certain types of fish live. Species A is used for a control group and Species B serves as the experimental group. A random sample of fish from both species are drawn and the following ages are recorded for life span in months.

Species A	Species B
100	102
90	100
36	99
82	12
91	80
56	64
101	79
77	

a. State the nondirectional alternative hypothesis.
b. State the null hypothesis
c. What is the tabled value of U for rejecting H_0 with $\alpha = .05_{2 \text{ tail}}$?
d. What is the value of U_{obt} for these data?
e. What do you conclude?

8. Someone has told you that left-handed people have different spatial reasoning abilities than right-handed people. You are skeptical, so you decide to test the idea. You randomly select 15 people from your class and administer a spatial reasoning test to them. A higher score reflects better spatial reasoning. You obtain the following results.

Left-handed	Right-handed
70	81
85	80
60	50
92	95
82	93
65	85
	90
	75
	84

a. State the null hypothesis.
b. State the alternative hypothesis.
c. What is your conclusion using $\alpha = .05_{2 \text{ tail}}$?
d. What type of error might you be making?
e. To what population do these results apply?

9. A psychologist wants to know if the nursing staff at a state mental hospital using a rating scale designed to assess interpersonal preferences rates patients with a diagnosis of schizophrenia any differently than patients diagnosed as having a major affective disorder. Since the psychologist is not certain about the scaling properties of

the rating scale she chooses to analyze the results with the Mann-Whitney U test. The data are shown below.

Schizophrenia Group	Major Affective Disorder Group
72	86
83	84
65	91
52	72
59	50
80	90
58	77
53	81
90	46
46	72

a. State H_1.
b. State H_0.
c. What do you conclude using $\alpha = .01_{2 \text{ tail}}$?
d. Give two possible explanations for the conclusion.

10. An educator believes that 9 year-old girls are more mature than 9 year-old boys. To test this hypothesis 11 boys and 12 girls are randomly selected from a 6th grade class at a neighborhood elementary school. Individually they listen to tape recordings of social situations and are then asked how they would respond. The answers are transcribed and later a panel of judges rate the responses and a "maturity score" is given to each response. The data are shown below. The higher the score the more mature the child was rated to be.

Girls	Boys
9.7	6.3
8.6	5.3
6.5	7.0
6.2	9.1
8.3	6.9
7.8	6.6
4.6	6.1
7.7	6.0
10.0	5.0
9.9	5.1
8.5	9.2
9.8	

    a. State H_1 using a directional hypothesis.
    b. State H_0.
    c. What would one conclude using $\alpha = .05_{1\ tail}$?

Answers: 1a. There is no value of U or U' that will allow rejection of H_0. **1b.** zero   **2.** CCCCEEEE or EEEECCCC   **3.** Increase the sample size.   **4.** 25   **5a.** .0286   **5b.** 0 **6a.** The overlap of distributions is maximum.   **6b.** Retain H_0.

7a. One species of fish has a different life span from the other.   **7b.** There is no difference in the life span of the two species of fish studied.   **7c.** Tabled U = 10; tabled U' = 46.   **7d.** $U_{obt} = 27.5$ **7e.** Retain H_0.

8a. There is no difference in performance on a spatial reasoning task between right and left handed people.   **8b.** There is a difference in performance on a spatial reasoning task between right and left handed people.   **8c.** $U_{obt} = 19.5$. Retain H_0, since tabled value for U is 10.   **8d.** Type II   **8e.** The people in your class.

9a. H_1: There is a difference between the ratings given by nurses on an interpersonal preference scale between patients with a diagnosis of schizophrenia and patients with a diagnosis of major affective disorder.   **9b.** H_0: There is no difference between the ratings given by nurses on an interpersonal preference scale between patients with a diagnosis of schizophrenia and patients with a diagnosis of major affective disorder.   **9c.** Tabled U = 16; $U_{obt} = 33$. Therefore, retain H_0.   **9d.** Either the power may be too low to detect a difference and the psychologist has made a Type II error or H_0 may in reality be true.

10a. H_1: Nine year-old girls are rated as significantly more mature than same aged boys as assessed by ratings of responses to social situations.   **10b.** H_0: There is no difference in maturity of nine year-old girls and boys or nine year-old girls may be less mature as assessed by ratings of responses to social situation.   **10c.** Tabled U = 38; $U_{obt} = 32$. Therefore, reject H_0.

TRUE-FALSE QUESTIONS

T F   1. In an independent groups design each subject serves as its own control.

T F   2. Any differences on the dependent variable in an independent groups design are due to the effects of the independent variable.

T F   3. Because the Mann-Whitney U test is more powerful than the sign test, if one obtains a nonsignificant result it is reasonable to accept the null hypothesis.

T F   4. The Mann-Whitney U test tests the difference between sample means.

T F   5. If chance alone were operating one would expect a great deal of overlap between two sets of scores.

T F   6. It is not possible to have $U_{obt} = 0$ if H_0 is true.

T F   7. The greater the degree of separation between the two groups of scores the higher the value for U'_{obt}.

T F   8. It is legitimate to use the "E's less than C's" method even when there are tied scores between groups.

T F   9. The value of Tabled U for $\alpha = .05_{2\ tail}$ where $n_1 = n_2 = 15$ is 64.

T F 10. $U + U' = R_1 R_2$.

T F 11. If $n_1 = n_2 = 5$, $U = 0$ and $U' = 20$ indicate the same degree of separation.

T F 12. If $n_1 = 10$ and $n_2 = 15$ then a value of $U' = 150$ indicates the greatest possible degree of separation.

T F 13. It is not possible to make a Type I error using the Mann-Whitney U test.

T F 14. To use the Mann-Whitney U test, n_1 must equal n_2.

T F 15. Unless there are tied ranks at the end, if the ranking is done correctly the last score will have a rank score equal to $n_1 + n_2$.

T F 16. One cannot properly apply the Mann-Whitney U test to nominal data.

T F 17. The null hypothesis for independent groups designs is that each sample is drawn from a different population.

Answers: 1. F   **2.** F   **3.** F   **4.** F   **5.** T   **6.** F   **7.** T   **8.** F **9.** T   **10.** F
11. F   **12.** T   **13.** F   **14.** F   **15.** T   **16** T   **17.** F.

SELF-QUIZ

1. An experimenter randomly divides a sample into two groups and gives each group a different treatment and analyzes the group scores. With what test(s) can the results be analyzed?

   a. sign test
   b. Mann-Whitney U test
   c. both of the above
   d. neither of the above

2. The statistics used for the Mann-Whitney U test measure _____.

   a. the differences between the two groups
   b. the direction of the differences between pairs of scores
   c. the power of the experiment
   d. the separation between the two sets of scores

3. The Mann-Whitney U test assumes that the population scores are normally distributed.

   a. true
   b. false

4. On which of the following types of scales can one use the Mann-Whitney U test?

   a. ordinal
   b. interval
   c. ratio
   d. all of the above

5. Consider the following set of scores: 81, 83, 84, 84, 87. What rank would you give to a score of 84?

   a. 3
   b. 3.5
   c. 4
   d. 4.5

6. Again consider the scores in question 5. What rank would you assign to the score of 87?

   a. 3.0
   b. 3.5
   c. 4.5
   d. 5.0

7. If $n_1 = 5$, $n_2 = 6$, and $U = U'$, what is the value of U and U'?

   a. 0
   b. 30
   c. 15
   d. 11

8. If there is only one way to get a value $U = 0$ by ordering C's and E's such that the C's' are lower than the E's and there are 20 possible orders of C s and E s altogether, what is the probability of getting $U = 0$ if chance alone is at work?

   a. .95
   b. .99
   c. .05
   d. .01

9. When using the Mann-Whitney U test, the probability of making a Type I error is determined by _____.

   a. alpha
   b. beta
   c. R_1
   d. R_2

10. In general, the sign test is _____ powerful than the Mann-Whitney U test.

   a. more
   b. less
   c. equally

11. The _____ test is used for repeated measures design.

   a. sign
   b. Mann-Whitney U
   c. both a and b
   d. neither a nor b

12. With the Mann-Whitney U test one does not have to be concerned with whether or not the samples are drawn randomly.

   a. true
   b. false

13. For the following data, the value of U_{obt} (not U'_{obt}) is _____ .

Control	Experimental
100	90
74	85
86	87
72	78
82	73
80	84
90	

    a. 14.5
    b. 20.5
    c. 21.5
    d. 27.5

Answers: 1. b   **2.** d   **3.** b   **4.** d   **5.** b   **6.** d   **7** . c   **8.** c   **9.** a   **10.** b
11. a   **12.** b   **13.** b.

13

SAMPLING DISTRIBUTIONS, SAMPLING DISTRIBUTION OF THE MEAN, THE NORMAL DEVIATE (z) TEST

CHAPTER OUTLINE

I. **Sampling Distribution of a Statistic**

A. <u>Definition:</u> a probability distribution of all the possible values of the statistic under the assumption that chance alone is operating.

B. Information given by sampling distributions.

   1. All the values that the statistic can take.

   2. The probability of getting each value under the assumption that it had resulted due to chance alone.

C. Examples of sampling distributions.

   1. Binomial distributions with $P = .50$ for the sign test in a replicated measures design.

   2. U or U' for the Mann-Whitney U test for the independent groups design.

D. Steps of data analysis.

   1. Calculate the appropriate statistic.

2. Evaluate the statistic based on its sampling distribution.

II. Generating Sampling Distributions

A. Derivation of sampling distribution of the statistic.

1. Determine all the possible different samples of size N that can be formed from the population.

2. Calculate the statistic for each of the samples.

3. Calculate the probability of getting each value of the statistic, if chance alone is operating.

B. Null Hypothesis Population.

1. Definition: actual or theoretical set of population scores which would result if the experiment were done on the entire population and the independent variable had no effect.

2. Use: to test the validity of the null hypothesis.

III. The Normal Deviate (z test)

A. Used when the parameters of the Null Hypothesis Population are known (μ and σ).

B. Uses the mean of the sample as a basic statistic to test the null hypothesis.

C. Must know the sampling distribution of the mean.

IV. Sampling Distribution of the Mean

A. Empirical approach.

1. Draw all possible different samples of a fixed size N from a specific population of raw scores having a mean μ and a standard deviation σ.

2. Calculate the mean of each sample.

3. Calculate the probability of getting each mean value if chance alone were operating.

B. Characteristics of the sampling distribution.

1. Resulting distribution is a distribution of sample means which itself has a mean and standard deviation.
2. Distribution of sample means is a population set of scores.

3. Mean of the sampling distribution of mean is signified by $\mu_{\overline{X}}$.

4. Standard deviation of the sampling distribution of mean is signified by $\sigma_{\overline{X}}$ and is also known as the standard error of the mean.

5. Mean of sampling distribution equals mean of raw score population:

$$\mu_{\overline{X}} = \mu$$

6. Standard deviation of the sampling distribution of mean is equal to the standard deviation of the raw score population divided by $\sqrt{N}$:

$$\sigma_{\overline{X}} = \sigma/\sqrt{N}$$

where   N = the size of the sample.

7. The sampling distribution is normally shaped depending on the shape of the raw score population and

   a. If the raw score distribution is normally shaped, the sampling distribution will be normally shaped.

   b. If the raw score distribution is not normally distributed, the sampling distribution of the mean will be

      i. Normal if N is greater than or equal to 300.

      ii. Usually normal for the behavioral sciences if $N \geq 30$.

V. Computation of z

A. Calculate sample mean, $\overline{X}$.

B. Use following formula:

$$z_{obt} = \frac{\overline{X}_{obt} - \mu_{\overline{X}}}{\sigma_{\overline{X}}}$$

where

$$\sigma_{\overline{X}} = \frac{\sigma}{\sqrt{N}}$$

VI. Evaluation of z and H_0

A. <u>Critical region for rejection of the null hypothesis:</u> the area under the curve which contains all the values of the statistic which allow rejection of H_0.

B. <u>Critical value of a statistic:</u> the value of the statistic which bounds the critical region.

C. Determine z_{crit} and assess whether z_{obt} falls within the critical region for rejection of H_0.

    1. Critical region of rejection is determined by the alpha (α) level.

    2. **If $| z_{obt} | \geq | z_{crit} |$, reject H_0. If not, retain H_0**

D. Conditions for use of the z test.

    1. Single sample is involved.

    2. μ and σ are known.

    3. Sampling distribution of mean must be normally distributed, i.e., $N \geq 30$, or the Null Hypothesis Population must be normally distributed.

VII.   Power and the z test

A. Main Points.

    1. Power is a measure of the sensitivity of the experiment to detect a real effect of the independent variable, if there is one.

    2. Defined as the probability of rejecting H_0, if the independent variable has a real effect.

    3. Varies directly with N and the magnitude of the independent variable's effect.

    4. Varies inversely with α.

    5. Power + Beta = 1.

B. Calculation of Power with N and μ_{real} given.

1. Determine the critical region for rejection of H_0, using $\overline{X}$ as the statistic.

2. Calculate the probability of getting a sample mean in the critical region for rejection of H_0, assuming sampling is random from the μ_{real} population

C. Determination of N to achieve a given level of power.
1. Equation:

$$N = \left[\frac{\sigma(z_{crit} - z_{obt})}{\mu_{real} - \mu_{null}} \right]^2$$

CONCEPT REVIEW

We have been considering the use of the (1) _____ method to investigate (2) _____. In analyzing our results we want to answer the question, what is the (3) _____ of getting the obtained results or results even more extreme if (4) _____ alone is responsible for the (5) _____ between the experimental and control scores.

Analyzing the results of an experiment involves two steps. First, calculating the appropriate (6) _____, and second, evaluating the statistic based on the appropriate (7) _____ (8) _____.

The sampling distribution of a statistic gives two important pieces of information. First, it gives all the (9) _____ that the (10) _____ can take. Second, it gives the (11) _____ of getting each value under the (12) _____ that it had resulted due to (13) _____ alone.

(1) scientific
(2) hypotheses
(3) probability
(4) chance
(5) difference
(6) statistic
(7) sampling
(8) distribution
(9) values
(10) statistic
(11) probability
(12) assumption
(13) chance

For example, with the (14) _____ measures design we used the sign test to analyze the data. The (15) _____ calculated was the (16) _____ of pluses in the sample of N (17) _____ scores. The results were analyzed using the (18) _____ (19) _____ with P = .50 as the appropriate (20) _____ (21) _____. In this case there is a (22) _____ sampling distribution with each different value of (23) _____.

So far, for the independent groups design we have used the (24) _____ test to analyze the data. With the protein and IQ experiment, for $n_1 = n_2 = 3$, we constructed a table to determine the (25) _____ of obtaining each possible value of U and U' This table was the (26) _____ (27) _____ of U for $n_1 = n_2 = 3$. Each value for n_1 and (28) _____ has its own (29) _____ distribution.

To draw our conclusions for any experiment we compare the (30) _____ of the (31) _____ statistic or one more extreme to the (32) _____ (33) _____. If the probability of the obtained statistic is (34) _____ than or (35) _____ to alpha we (36) _____ the (37) _____ (38) _____. If not, we (39) _____ H_0. This strategy is the same for all experiments. What varies is the (40) _____ used and its accompanying (41) _____ (42) _____.

Using the empirical approach, sampling distributions of a statistic are derived by first determining all the (43) _____

(14) replicated
(15) statistic
(16) number
(17) difference
(18) binomial
(19) distribution
(20) sampling
(21) distribution
(22) different
(23) N

(24) Mann-Whitney U

(25) probability

(26) sampling
(27) distributionl
(28) n_2
(29) sampling

(30) probability
(31) obtained
(32) alpha
(33) level
(34) less
(35) equal
(36) reject
(37) null
(38) hypothesis
(39) retain
(40) statistic

(41) sampling.
(42) distribution

(43) possible

different samples of size (44) _____ that can be formed from the

population; second, calculating the (45) _____ for each of these

(46) _____; and third, calculating the (47) _____ of getting

each value if (48) _____ alone is operating. This is done using

a theoretical set of scores which would result if the independent

variable had no effect. This population set of scores is called
the (49) _____ (50) _____ (51) _____. More formally,

the Null Hypothesis Population is an actual or theoretical set

of population scores which would result if the experiment were

done on the entire (52) _____ and the (53) _____ variable

had (54) _____ effect. It is called the Null Hypothesis Population

because it is used to test the (55) _____ of the (56) _____

(57) _____. A sampling distribution gives all the values a

statistic can take, along with the probability of getting each value

if (58) _____ is (59) _____ from the Null Hypothesis

Population.

(44) N

(45) statistic

(46) samples
(47) probability
(48) chance

(49) Null
(50) Hypothesis
(51) Population

(52) population
(53) independent
(54) no

(55) validity
(56) null
(57) hypothesis

(58) sampling
(59) random

The Normal Deviate

The (60) _____ test or (61) _____ (62) _____ test is

used when we know the (63) _____ of the Null Hypothesis

Population. The statistic used is the (64) _____ of the sample.

Therefore, the appropriate probability distribution to use is the

sampling distribution of the (65) _____. This sampling dis-

tribution is constructed by taking a specific population of

(66) _____ scores having a mean of (67) _____ and

(60) z
(61) normal
(62) deviate
(63) parameters
(64) mean

(65) mean

(66) raw

(67) μ

standard deviation of (68) _____. Then one draws all possible

different (69) _____ of size (70) _____, calculates the

(71) _____ of each sample, and calculates the (72) _____

of getting each mean value if chance alone were operating. The

result is the sampling distribution of the mean for samples of size

(73) _____ for the specific population parameters

(74) _____ and (75) _____.

  The sampling distribution of the mean for a given size N has

features of its own, namely a (76) _____ and (77) _____

(78) _____. The mean of the sampling distribution of the

mean is symbolized by (79) _____. The standard deviation

of the sampling distribution of the mean is symbolized by

(80) _____. This standard deviation is also known as the

(81) _____ (82) _____ of the mean. The sampling distribution

of the mean has a mean equal to the (83) _____ of the

(84) _____score population. In equation form:

$$\mu_{\overline{X}} = \mu$$

It also has a standard deviation equal to the (85) _____

(86) _____ of the (87) _____

score population divided by (88) _____. In equation form:

$$(89) \text{ _____} = \sigma / \sqrt{N}$$

(68) σ

(69) samples
(70) N
(71) mean
(72) probability

(73) N

(74) μ
(75) σ

(76) mean
(77) standard
(78) deviation

(79) $\mu_{\overline{X}}$

(80) $\sigma_{\overline{X}}$

(81) standard
(82) error
(83) mean

(84) raw

(85) standard

(86) deviation
(87) raw
(88) $\sqrt{N}$

(89) $\sigma_{\overline{X}}$

The sampling distribution of the mean is (90) _____ shaped (90) normally

depending on the raw score population and on the (91) _____ (91) sample

(92) _____. If the shape of the raw score population is normally (92) size

distributed, then the sampling distribution will be (93) _____ (93) normally

distributed (94) _____ of sample size. If the raw scores are not (94) regardless

normally distributed, then the sampling distribution of the mean

becomes more (95) _____ as N (96) _____. If N is greater (95) normal
(96) increases
than (97) _____ then the sampling distribution is almost (97) 300

(98) _____ normally shaped. In the behavioral sciences (98) always

if N is greater than or equal to (99) _____, it is usually assumed (99) 30

that the sampling distribution of the mean is normally shaped.

The z equation for sample means is very similar to the z

equation for single scores. The equation for the z test for sample

means is:

$$z_{obt} = \frac{(100) \underline{\quad} - (101) \underline{\quad}}{(102) \underline{\quad}}$$

(100) $\overline{X}_{obt}$
(101) $\mu_{\overline{X}}$ or μ
(102) $\sigma_{\overline{X}}$

One can then compare the (103) _____ of getting (104) _____ (103) probability
(104) z_{obt}
with the (105) _____ (106) _____ to draw conclusions about (105) alpha
(106) level
the experiment. Another way to analyze the results is to determine

the (107) _____ (108) _____ for rejection of the null (107) critical
(108) region
hypothesis. This is the (109) _____ under the curve which (109) area

contains all the (110) _____ of the (111) _____ which will (110) values
(111) statistic

allow rejection of the null hypothesis.

The (112) _____ (113) _____ of a statistic is the value   (112) critical
                                                 (113) value

of the statistic which (114) _____ the critical region. The   (114) bounds

critical region is determined by the (115) _____ (116) _____.   (115) alpha
                                                 (116) level

The combined area under the critical region must (117) _____.   (117) equal

the alpha level. If | (118) _____ | $\geq$ | (119) _____ |, reject   (118) z_{obt}
                                                 (119) z_{crit}

the null hypothesis. If not, (120) _____ the null hypothesis. In   (120) retain

order to use the z test there must be a (121) _____ sample   (121) single

and the parameters of the Null Hypothesis Population,

(122) _____ and (123) _____ must be known.   (122) μ
                                                 (123) σ

Also, the sampling distribution of the mean must be

(124) _____ (125) _____. This means N must be greater   (124) normally
                                                 (125) distributed

than or equal to (126) _____ or the Null Hypothesis Population   (126) 30

must be normally distributed.

The power of a test or of an experiment is a measure of the

(127) _____ of the experiment to the real effect of the independ-   (127) sensitivity

ent variable. Thus, power is defined as the probability of

(128) _____ H_0 when the independent variable has a real   (128) rejecting

effect in the experiment. It is also the probability of rejecting H_0

when (129) _____ is false. If the real effect of the independent   (129) H_0

variable increases, power (130) _____. If the number of subjects   (130) increases

in an experiment is increased, with other factors held constant,

power also (131) _____. There is an (132) _____ relation-   (131) increases
                                                 (132) inverse

ship between power and beta. The (133) _____ the power,   (133) higher

the less the chance of a Type II error. Since making alpha more stringent, causes beta to (134) _____, making alpha more   (134) increase

stringent, will cause power to (135) _____. In order to minim-   (135) decrease

ize Type I and II errors, it is desirable to make alpha (136) _____   (136) low

and power (137) _____. For a given low (stringent) alpha level,   (137) high

power can be increased by increasing (138) _____ and   (138) N

increasing the (139) _____of the (140) _____   (139) effect

(141) _____.   (140) independent
                                                     (141) variable

EXERCISES

For the following problems assume the Null Hypothesis Populations are normally distributed.

1. What is the critical value for z for each of the following alpha levels?

   a. $.05_{2 \text{ tail}}$
   b. $.01_{2 \text{ tail}}$
   c. $.05_{1 \text{ tail}}$
   d. $.01_{1 \text{ tail}}$
   e. $.02_{2 \text{ tail}}$

2. What is the value of $\mu_{\overline{X}}$ for the following values of μ?

   a. 15.2
   b. 14.0
   c. 10.0

3. List all the possible values of $\overline{X}$ for samples of size N = 2 taken from the population of scores 0, 1, 2 (with replacement).

4. For the means of size N = 2 in problem 3, what is the value for:

   a. $\mu_{\overline{X}}$
   b. $\sigma_{\overline{X}}$

5. Assume the following values of σ have been obtained for different Null Hypothesis Populations. Calculate the value of the standard error of the mean for samples of size N taken from the respective populations.

    a. $\sigma = 26.2$, $N = 9$

    b. $\sigma = 5.6$, $N = 2$

    c. $\sigma = 13.0$, $N = 6$

    d. $\sigma = 1000$, $N = 100$

    e. $\sigma = 10.0$, $N = 3$

6. In a population with $\mu = 100$ and $\sigma = 15$, what is the probability if we randomly drew a sample of $N = 9$ we would get a value of $\overline{X} \geq 108$?

7. In problem 6, if you were testing the hypothesis that the sample mean (108) was drawn from the Null Hypothesis Population with $\mu = 100$ and $\sigma = 15$, what would you conclude using $\alpha = .05_{1\ tail}$?

8. You have just read that chickens have a relatively low level of body fat. The population mean percentage of fat in a chicken is reported to be 32% with the standard deviation of 8.7%. How likely is it that you could randomly select a sample of 12 chickens from the population which would have a mean of 26% body fat or less?

9. A computer memory manufacturer specifies that its memory chip stores data incorrectly an average of 6.3 out of 10 million cycles with a standard deviation of 0.48. A batch of 30 chips your company ordered stores data incorrectly an average of 6.9 times per 10 million cycles.

    a. Does it seem reasonable that your 30 chips are a random sample from a population with the specifications given by the computer memory manufacturer? Use $\alpha = .05_{2\ tail}$.

    b. What is the critical value of z?

10. The average weight of a species of laboratory rats at birth is 27.6 grams with a standard deviation of 3.4 grams. You wish to test the hypothesis that a different maternal feeding cycle but the same quantity and quality will affect birth-weight. A sample of 10 female rats given the new feeding schedule gave birth to 56 baby rat pups with an average weight of 25.9 grams.

    a. What do you conclude using $\alpha = .01_{2\ tail}$?

    b. If the real effect of the new feeding schedule is to cause an average weight at birth of 24.5 grams, what is the power of the experiment to detect this effect?

    c. If the sample size was increased to 25, what is the power of the experiment to detect the real effect given in 10b?

    d. What size of sample would it be necessary to use to have power =.9000 to detect an effect which produces an average birth weight of 26.0?

11. From a population of 4 different scores (e.g. 3, 5, 7, 9) how many samples of size $N = 2$ are possible sampling one at a time with replacement? (Solve by enumeration.)

12. The average life of a light bulb is 862 hours with a standard deviation of 51 hours. A new manufacturing process results in a sample of 24 bulbs with a mean life of 899 hours until burn out.

    a. Would you recommend a change in manufacturing technique using $\alpha = .01_{1 \text{ tail}}$?

    b. What type error might you be making?

    c. If the new manufacturing process has an effect so as to produce an average life of 880 hours, what is the power of the experiment to detect this effect?

    d. What should the sample size be to have a power =.8000 to detect the real effect postulated in 12c?

    e. What effect does making alpha less stringent have on power? Illustrate by recomputing power for question 12c, only this time use $\alpha = .05_{1 \text{ tail}}$.

13. Using $\alpha = .01_{2 \text{ tail}}$, test the hypothesis that the following sample could have been drawn from a population with $\mu = 27.2$ and $\sigma = 2.1$.

$$X: \quad 25.2, \ 29.9, \ 24.8, \ 26.0, \ 22.1$$

Answers: 1a. ±1.96   **1b.** ±2.58   **1c.** 1.64(5)   **1d.** 2.33   **1e.** ±2.33   **2a.** 15.2 **2b.** 14.0   **2c.** 10.0   **3.** 0, .5, 1, .5, 1, 1.5, 1 1.5, 2   **4a.** 1.0   **4b.** .5773   **5a.** 8.73 **5b.** 3.96   **5c.** 5.31   **5d.** 100.00   **5e.** 5.77   **6.** .0548   **7.** Retain H_0 **8.** p = .0084 **9a.** $z_{obt} = 6.85$, therefore, reject H_0. Conclude that the sample of 30 chips with a mean of 6.9 failures could not reasonably have been drawn from the Null Hypothesis Population. **9b.** ±1.96   **10a.** $z_{obt} = -1.58$ and $z_{crit} = ±2.58$. Retain H_0.   **10b.** Power = .6179 **10c.** Power = .9761   **10d.** N = 67   **11.** 16   **12a.** Reject H_0 (z_{obt}= 3.55, z_{crit} = 2.33). Assuming there were no economic disadvantages in changing manufacturing processes, it appears as if this process should be adopted.   **12b.** Type I   **12c.** Power = .2743 **12d.** N = 81   **12e.** It increases power. Power = .5319   **13.** Retain H_0 ($z_{obt} = -1.70$, $z_{crit} = ±2.58$).

TRUE FALSE QUESTIONS

T F   1. The binomial distribution is an example of a sampling distribution.

T F   2. A sampling distribution is based on the assumption that the independent variable has an effect on the dependent variable.

T F   3. Only one sampling distribution can apply to any given problem or experiment.

T F   4. In the case of the Mann-Whitney U test, if the independent variable had no effect, the Null Hypothesis Population would have U_{obt} approximately equal to U_{crit}.

T F   5. The sampling distribution of the mean changes as sample size changes.

T F   6. A sampling distribution gives one the values a statistic can take, along with the probability of getting each value if sampling is random from the H_1 population.

T F   7. To use the normal deviate one must know the population parameters of the Null Hypothesis Population.

T F   8. If the sample mean is different from the population mean, H_0 must be false.

T F   9. $\sigma_{\overline{X}}$ is sometimes called the standard error of the mean because each sample can be considered an estimate of the mean of the raw score population and variability between sample means occur due to errors in estimation.

T F   10. For sampling distribution of the mean, $\mu_{\overline{X}} = \mu$.

T F   11. As N increases, $\sigma_{\overline{X}}$ decreases.

T F   12. In general, if $N \geq 30$, the sampling distribution of the mean will be normally shaped.

T F   13. For the population 2, 4, 6, 8, 10; the value of $\mu_{\overline{X}} = 10/\sqrt{5}$

T F   14. The formula for z_{obt} for the sampling distribution of the mean is $z = (X - \mu)/\sigma$.

T F   15. If z_{obt} falls within the critical region for rejection of H_0, one can be certain H_0 is in reality false.

T F   16. If $\alpha = .05_{2\ tail}$, the critical value of z is ± 1.96.

T F   17. The z test is used to test whether or not it is reasonable to assume a sample of size N was likely to have been drawn from a population with known parameters μ and σ.

T F   18. For the sampling distribution of the mean, $\sigma_{\overline{X}} = \sigma$.

T F   19 $\alpha + \beta = 1.00$.

T F   20   If there is a real effect, the higher the power, the more likely H_0 will be rejected.

T F   21   Increasing N usually results in an increase of power.

T F   22   If H_0 is false, but power is low, there is a high probability we will err in our conclusion.

T F 23 By using a stringent alpha level, and designing the experiment for high power, we maximize the probability of correctly concluding regardless of whether H_0 is true or false.

T F 24 Power is independent of the magnitude of the independent variable's effect.

T F 25 Power is an esoteric topic that has very little practical utility.

Answers: 1. T   **2.** F   **3.** F   **4.** F   **5.** T   **6.** F   **7.** T   **8.** F   **9.** T   **10.** T
11. T   **12.** T   **13.** F   **14.** F   **15.** F   **16.** T   **17.** T   **18.** F   **19.** F   **20.** T
21. T   **22.** T   **23.** T   **24.** F   **25.** F.

SELF-QUIZ

1. A raw score distribution that is negatively skewed will produce a sampling distribution of the mean for N = 4 which is _____.

   a. also negatively skewed
   b. normally distributed
   c. positively skewed
   d. cannot be determined

2. A raw score distribution which has a moderate negative skew will result in a sampling distribution of the mean for N = 42 which is _____.

   a. also negatively distributed
   b. normally distributed
   c. positively skewed
   d. cannot be determined

3. If one draws all possible samples for various values of N from the same population of raw scores, as N increases _____.

   a. the standard error of the mean increases
   b. the standard error of the mean stays the same
   c. the standard error of the mean decreases
   d. the standard error of the mean cannot be calculated

4. If one draws all possible samples for various values of N from the same population of raw scores, as N increases _____.

    a. the mean of the sampling distribution of the mean increases
    b. the mean of the sampling distribution of the mean stays the same
    c. the mean of the sampling distribution of the mean decreases
    d. none of the above

5. In cases where N > 1, the relationship between the raw score population standard deviation and the standard error is _____.

    a. the standard error is greater than the standard deviation
    b. the standard error is less than the standard deviation
    c. the standard error equals the standard deviation
    d. the standard error is the standard deviation

6. The area under the critical region for rejection must _____.

    a. equal alpha
    b. be less than alpha
    c. be greater than alpha
    d. equal one-half of alpha

7. The critical value of the statistic _____.

    a. is independent of alpha
    b. is equal to alpha
    c. is always the same in the z distribution
    d. depends on alpha

8. The critical value for the z distribution using $\alpha = .05_{2 \text{ tail}}$ is _____.

    a. ±1.96
    b. ±2.58
    c. ±1.64
    d. ±2.33

9. The critical value for the z distribution using $\alpha = .01_{2 \text{ tail}}$ is _____.

    a. ±1.96
    b. ±2.58
    c. ±1.64
    d. ±2.33

10. Assuming that the population mean is 47.2 and the population deviation is 6.4, what is the z_{obt} value for a sample mean of 52.1 if N = 8?

    a. 1.96
    b. 2.17
    c. 1.73
    d. 0.77

11. If z_{obt} was 2.60 and z_{crit} was 2.58, one would _____.

    a. accept H_0
    b. reject the alternative hypothesis
    c. retain H_0
    d. reject H_0

12. To use the z test the data _____.

    a. must be normally distributed
    b. must be sampled from a normal distribution
    c. must be sampled from a normal distribution if sample size is 10
    d. can be any shape if N = 10

13. What would one conclude about the null hypothesis that a sample of N = 46 with a mean of $\overline{X}$ = 104 could reasonably have been drawn from a population with the parameters of μ = 100 and σ = 8? Use α = .05$_{2\ tail}$.

    a. accept H_0
    b. reject H_1
    c. reject H_0
    d. retain H_0

14. If the population standard deviation is 19.0 and the sample size is 19, then the standard error equals _____.

    a. 19.00
    b. 1.00
    c. 4.36
    d. cannot be determined; need to know population mean

15. If one draws two samples of size N from a distribution of raw scores (with replacement), then the means of the two samples will _____.

    a. be equal
    b. perhaps be equal but not necessarily
    c. be equal to the standard error
    d. be equal to 1.0

16. What we are testing with the z test in this chapter is _____.

    a. the sample mean
    b. the population mean
    c. the sample error
    d. the standard error

17. Could the following sample reasonably have been drawn from a normal population with a mean of 20 and standard deviation of 1.5 using $\alpha = .01_{2\ tail}$?

    X: 21, 21, 21, 20, 22, 20, 22?

    a. yes
    b. no
    c. cannot be tested with z test
    d. insufficient information

18. A health psychologist knows the smoking population at the hospital where she works smokes an average of 18 cigarettes per day with a standard deviation of 7. She plans to conduct a program to reduce smoking. If she has 25 persons in her program, what is the power to detect a real effect of the program such that cigarette consumption is reduced, on the average, by 5 cigarettes? Assume population normality and use $\alpha = .05_{1\ tail}$.

    a. Insufficient information
    b. .9732
    c. .9463
    d. .8925

Answers: 1. a **2.** b   **3.** c   **4.** b   **5.** b   **6.** a   **7.** d   **8.** a   **9.** b   **10.** b   **11.** d   **12.** c   **13.** c   **14.** c   **15.** b   **16.** a   **17.** a   **18.** b

14 | STUDENT'S t TEST FOR SINGLE SAMPLES

CHAPTER OUTLINE

I. Introduction

A. Use of the t test indicated when:

    1. The mean of the Null Hypothesis Population can be specified

    2. Standard deviation is unknown (which also means the z test cannot be used)

B. Uses covered in this chapter:

    1. Analysis of data in experiments with a single sample

    2. Determining the significance of Pearson r.

II. Comparison of z and t Tests

A. <u>Formulae:</u>

$$z_{obt} = \frac{\overline{X}_{obt} - \mu}{\sigma_{\overline{X}}} \quad \text{where} \quad \sigma_{\overline{X}} = \frac{\sigma}{\sqrt{N}}$$

$$t_{obt} = \frac{\overline{X}_{obt} - \mu}{s_{\overline{X}}} \quad \text{where} \quad s_{\overline{X}} = \frac{s}{\sqrt{N}}$$

B. Difference between equations is that when σ is unknown we estimate it using s, the sample standard deviation. When this is done the resulting test is called the t test.

C. $s_{\overline{X}}$ is called the estimated standard error of the mean.

III. Sampling Distribution of t

A. <u>Definition:</u> the sampling distribution of t is a probability distribution of the t values which would occur if all possible different samples of a fixed size N were drawn from the Null Hypothesis Population. It gives (1) all the possible t values for samples of size N and (2) the probability of getting each value if sampling is random from the Null Hypothesis Population.

B. Characteristics.

1. Family of curves: many curves with a different curve for each sample size N.

2. Shaped similarly to z distribution if:

a. sample size $\geq$ 30

b. H_0 population is normally distributed.

C. Degrees of freedom (df). The number of scores that are free to vary.

$$\mathbf{df = N - 1}$$

D. Comparison of z and t distribution.

1. t and z are both symmetrical about 0

2. As df increases t becomes more similar to z

3. As df approaches ∞, t becomes identical to z

4. At any value of df < ∞, the t distribution has more extreme values than z (i.e., tails of the t distribution are more elevated than in the z distribution)

5. For a given alpha level, $t_{crit} > z_{crit}$

IV. Calculations and Use of t

A. Calculation of t from raw scores:

$$t_{obt} = \frac{\overline{X}_{obt} - \mu}{\sqrt{\dfrac{SS}{N(N-1)}}}$$

B. Appropriate use of t requires that sampling distribution of $\overline{X}$ is normal. This can result if N ≥ 30 or population of raw scores is normal.

V. Confidence Intervals for the Population Mean

A. <u>Definition:</u> A confidence interval is a range of values which probably contains the population mean. Confidence limits are the values that bound the confidence interval. Example: The 95% confidence interval is an interval such that the probability is .95 that the interval contains the population value.

B. Formula for confidence interval:

$$\mu_{lower} = X_{obt} - (s_{\overline{X}})t_{crit}$$

$$\mu_{upper} = X_{obt} + (s_{\overline{X}})t_{crit}$$

VI. Testing Significance of Pearson r.

A. <u>Rho (ρ):</u> this is the Greek letter to symbolize the population correlation coefficient.

B. Nondirectional alternative hypothesis asserts ρ ≠ 0.

C. Directional alternative hypothesis asserts that ρ is positive or negative depending on the predicted direction of the relationship.

D. Sampling distribution of r can be generated by taking all samples of size N from a population in which $\rho = 0$ and calculating r for each sample. By systematically varying the population scores and N, the sampling distribution of r is generated.

E. Using t test to evaluate significance of r:

$$t_{obt} = \frac{r_{obt} - \rho}{s_r} \quad \text{where} \quad s_r = \sqrt{\frac{1 - (r_{obt})^2}{N - 2}}$$

with df = N - 2 and N equals the number of pairs of X, Y scores.

F. Using r_{crit} to evaluate the significance of r:

$$\text{if } |r_{obt}| \geq |r_{crit}|, \text{ reject } H_0$$

The values of r_{crit} can be calculated directly. Values of r_{crit} are shown in Table E.

CONCEPT REVIEW

The (1) _____ test is appropriate in situations where both

the (2) _____ and (3) _____ of the Null Hypothesis Popula-

tion are known. However, it is more common that we can specify

the mean of the Null Hypothesis Population and the standard

deviation is unknown. In this case we can use the (4) _____ test.

The formula for the t test is:

$$t_{obt} = \frac{(5) \underline{\quad} - (6) \underline{\quad}}{(7) \underline{\quad}}$$

(1) z

(2) mean
(3) standard
      deviation

(4) t

(5) $\overline{X}$
(6) μ
(7) $s_{\overline{X}}$

where

$$s\overline{x} = (8) \underline{\hspace{2cm}} / \sqrt{(9) \underline{\hspace{2cm}}}$$

(8) s
(9) N

This equation differs from the equation for z_{obt} only in that

(10) \underline{\hspace{2cm}} is used instead of (11) \underline{\hspace{2cm}} in the denom-

(10) s
(11) σ

inator. When σ is unknown, we (12) \underline{\hspace{2cm}} it using the estimate

(12) estimate

given by (13) \underline{\hspace{2cm}}. When σ is estimated, the resulting

(13) s

statistic is called (14) \underline{\hspace{2cm}}. The (15) \underline{\hspace{2cm}}

(14) t
(15) sampling

(16) \underline{\hspace{2cm}} of t is a probability distribution of the t values

(16) distribution

which would occur if all possible different samples of a fixed size

were drawn from the (17) \underline{\hspace{2cm}} (18) \underline{\hspace{2cm}} (19) \underline{\hspace{2cm}}.

(17) Null
(18) Hypothesis
(19) Population

It gives all the possible values of (20) \underline{\hspace{2cm}} for samples of

(20) t

size N and the (21) \underline{\hspace{2cm}} of getting each value if sampling is

(21) probability

(22) \underline{\hspace{2cm}} from the Null Hypothesis Population. If the Null

(22) random

Hypothesis Population is (23) \underline{\hspace{2cm}} (24) \underline{\hspace{2cm}} or N

(23) normally
(24) shaped

(25) \underline{\hspace{2cm}} 30, then the t distribution looks very much like

(25) $\geq$

the (26) \underline{\hspace{2cm}} distribution, except that there is a

(26) z

(27) \underline{\hspace{2cm}} of curves that vary with (28) \underline{\hspace{2cm}} (29) \underline{\hspace{2cm}}.

(27) family
(28) sample
(29) size

The t distribution varies uniquely with (30) \underline{\hspace{2cm}} of

(30) degrees

(31) \underline{\hspace{2cm}}. The degrees of freedom, symbolized

(31) freedom

(32) \underline{\hspace{2cm}}, for any statistic is the (33) \underline{\hspace{2cm}} of scores that

(32) df
(33) number

are free to (34) \underline{\hspace{2cm}} in calculating that (35) \underline{\hspace{2cm}}. For the

(34) vary
(35) statistic

t test for single samples there are (36) \underline{\hspace{2cm}} df.

(36) N - 1

The t distribution is (37) \underline{\hspace{2cm}} about zero. When df

(37) symmetrical

approaches (38) _____ the t distribution becomes (39) _____   (38) infinity
(39) identical

to the z distribution. At any number of df less than infinity the

distribution has (40) _____ variability than the z distribution.   (40) more

That is, the (41) _____ of the t distribution is higher than the z   (41) tail

distribution.   For a given alpha level, the critical value of t is

(42) _____ than for z. This is because we use (43) _____ to   (42) higher
(43) s

estimate (44) _____ and we cannot be certain that our estimate   (44) σ

is perfectly accurate.

    To calculate the value of t_{obt} from raw data we can use the

following formula.

$$t_{obt} = \frac{\overline{X}_{obt} - \mu}{\underset{(46)\ \rule{1cm}{0.4pt}}{}\left(\underset{(47)\ \rule{1cm}{0.4pt}}{} - \underset{(48)\ \rule{1cm}{0.4pt}}{}\right)}$$

(45) SS
(46) N
(47) N
(48) 1

with (45) _____ in the numerator denominator position

Where   **(49)** _____ $= \Sigma\ X^2 - (\Sigma\ X)^2/N$ (49) SS

To use the t test, the sampling distribution of (50) _____ must   (50) $\overline{X}$

be (51) _____.   (51) normal

Confidence Intervals

    Unless we have sampled the (52) _____ population we   (52) entire

cannot know the population mean exactly. We use the

(53) _____ mean as an (54) _____ of the true population   (53) sample
(54) estimate

mean. When only one value is used as an estimate this is

called a (55) _____ (56) _____. In estimating the population   (55) point
(56) estimate

mean, the usual way is to give a (57) _____ of values for

which one is reasonably (58) _____ that the range

(59) _____ the population mean. This is called (60) _____

(61) _____. A (62) _____ interval is a range of (63) _____

which probably contains the population mean. The confidence

(64) _____ are the values that (65) _____ the confidence

interval. The 95% confidence interval is an interval such that the

(66) _____ is (67) _____ that the interval contains the pop-

ulation value. There is a probability of (68) _____ that such an

interval will not contain the population value. The general formula

for the confidence interval is

$$\mu_{\text{lower}} = (69) \,\rule{1cm}{0.4pt}\, - (70) \,\rule{1cm}{0.4pt}\, \times (71) \,\rule{1cm}{0.4pt}\,$$

$$\mu_{\text{upper}} = (72) \,\rule{1cm}{0.4pt}\, + (73) \,\rule{1cm}{0.4pt}\, \times (74) \,\rule{1cm}{0.4pt}\,$$

where t_{crit} = the critical (75) _____ - tailed value of t correspond-

ing to the desired confidence interval.

The (76) _____ confidence interval will always be larger

(wider) than the (77) _____ confidence interval. The greater

the interval, the more (78) _____ we have that it contains

the population mean.

Significance of r

To determine whether a correlation exists in the population,

we must test the (79) _____ of the obtained (80) _____.

The population correlation coefficient is symbolized by the

(57) range

(58) confident

(59) includes
(60) interval
(61) estimation
(62) confidence
(63) values

(64) limits
(65) bound

(66) probability
(67) .95
(68) .05

(69) $\overline{X}_{\text{obt}}$
(70) $s_{\overline{X}}$
(71) t_{crit}
(72) $\overline{X}_{\text{obt}}$
(73) $s_{\overline{X}}$
(74) t_{crit}

(75) one

(76) 99%

(77) 95%

(78) confidence

(79) significance
(80) r

Greek letter (81) _____. The nondirectional alternative hypo-

thesis asserts that ρ = (82) _____. The null hypothesis is tested

by assuming the sample set of X and Y scores, which have a

correlation equal to (83) _____, is drawn from a population

where (84) _____ = 0. Using the t test, the significance of r

can be evaluated using the formula:

$$t_{obt} = \frac{(85) _____ - \rho}{(86) _____}$$

where

$$s_r = \sqrt{\frac{1- (87) _____}{(88) _____ - 2}}$$

with df = (89) _____ - (90) _____. N is the number of

(91) _____ of scores and s_r is the (92) _____ of the

standard deviation of the sampling distribution of (93) _____

Again, this is a t test because we are (94) _____ the standard

deviation of the (95) _____ (96) _____. A table of r_{crit} can

be made by substituting (97) _____ into the above equation

and solving for (98) _____ for any level of (99) _____

and (100) _____. Then we can apply the decision rule:

If I(101) _____ I ≥ I(102) _____ I, reject H$_0$

(81) ρ(rho)

(82) 0

(83) r_{obt}

(84) ρ

(85) r_{obt}

(86) s_r

(87) r_{obt}

(88) N

(89) N
(90) 2
(91) pairs
(92) estimate
(93) r

(94) estimating

(95) sampling
(96) distribution
(97) t_{crit}

(98) r_{crit}
(99) α
(100) df

(101) r_{obt}
(102) r_{crit}

EXERCISES

For the following problems, assume the Null Hypothesis Populations are normally distributed.

1. What are the values for the degrees of freedom for samples of the following sizes (assuming one is calculating a t test for a single sample)?

   a. 25
   b. 24
   c. 2
   d. 11
   e. 9

2. A sample has a value of $\overline{X}_{obt} = 46$, s = 8, and N = 12.

   a. What is the value of t_{obt} to test the hypothesis that this sample could reasonably have been drawn from the Null Hypothesis Population with $\mu = 50$?
   b. What is the critical value of t using $\alpha = .05_{2\ tail}$?
   c. What do you conclude?

3. What are the critical values of t for each of the following values of N and alpha using a nondirectional hypothesis?

N	α
a. 12	.05
b. 20	.01
c. 2	.05
d. 5	.02
e. 19	.01

   Now using a directional hypothesis?

N	α
f. 13	.025
g. 17	.005
h. 8	.05
i. 15	.01
j. 10	.05

4. One wishes to investigate the hypothesis that exercise reduces systolic blood pressure. The population mean for the systolic blood pressure of people who do not exercise is 120 mmHg. A sample of 13 members of a local running club have a mean systolic blood pressure of 113 mmHg with a standard deviation of 9.2 mmHg.

    a. State H_1 (nondirectional)
    b. State H_0
    c. What do you conclude using $\alpha = .01_{2 \text{ tail}}$

5. Using $\alpha = .05_{2 \text{ tail}}$, is it reasonable to assume that the following set of scores could have been randomly drawn from a population with a mean of 74.5?

    X: 73.0, 72.1, 72.0, 69.1, 70.8

6. The average growth per day of a certain type lawn seed using standard fertilizer is 2.6 inches per week. You wish to test the effects of a fertilizer on growth rate. You add fertilizer to several different patches of grass grown from that seed and observe the following results:

Growth per Week (inches)
1.9
2.0
3.5
4.2
4.0
3.2
2.8

    a. What is the value of t_{obt}?
    b. What do you conclude using $\alpha = .05_{2 \text{ tail}}$?

7. Consider the following set of scores:

    102, 102, 106, 105, 104, 104, 107, 108

    a. What is the 95% confidence interval for the population mean?
    b. What is the 99% confidence interval for the population mean?

8. If $\overline{X} = 50$ and $s = 6.7$:

    a. If N = 18, what is the 95% confidence interval for the population mean?
    b. If N = 18, what is the 99% confidence interval for the population mean?
    c. If N = 10, what is the 95% confidence interval for the population mean?
    d. If N = 10, what is the 99% confidence interval for the population mean?

9. Assume that $\mu = 500$ and $\sigma = 100$. Your study shows a sample of size 22 with a mean of 530 and standard deviation of 113.

   a. What is the most powerful test to use to test the hypothesis that the mean of the sample was drawn from the above Null Hypothesis Population?

   b. What is the value of the test statistic?
   c. What do you conclude using $\alpha = .05_{2\ tail}$?

10. Using the information in problem 9:

   a. What is the value of t_{obt}?
   b. Does z or t allow one to reject H_0 more easily?
   c. What type error might one be making in this problem?

11. Assume that you have just calculated the correlation coefficient from an experiment with N = 16 pairs of observations. The value of r_{obt} is .493.

   a. What is the value of t_{obt} for testing the hypothesis that r is significantly different from zero?
   b. What do you conclude using $\alpha = .05_{2\ tail}$?
   c. What is the value of r_{crit}?

12. Consider the following set of scores:

X	10	15	20	25	30	35
Y	11	12	16	13	19	20

   a. What is the value of the correlation coefficient?
   b. What is the value of r_{crit} using $\alpha = .05_{2\ tail}$?

   c. Is r_{obt} significantly different from zero using $\alpha = .05_{2\ tail}$?

Answers: 1a. 24   **1b.** 23   **1c.** 1   **1d.** 10   **1e.** 8   **2a.** -1.73   **2b.** ±2.201   **2c.** retain H_0 **3a.** ±2.201   **3b.** ±2.861   **3c.** ±12.706   **3d.** ±3.747   **3e.** ±2.878   **3f.** 2.179   **3g.** 2.921   **3h.** 1.895   **3i.** 2.624   **3j.** 1.833   **4a.** exercise affects systolic blood pressure   **4b.** exercise has no effect on systolic blood pressure **4c.** $t_{obt} = -2.74$ while $t_{crit} = ±3.055$; therefore retain H_0 **5.** Since $t_{obt} = -4.61$ and $t_{crit} = ±2.776$, reject H_0, i.e. it is not reasonable   **6a.** $t_{obt} = 1.42$ **6b.** retain H_0 since $t_{crit} = ±2.447$ **7a.** 102.92 to 106.58   **7b.** 102.04 to 107.46; using s = 2.188   **8a.** 46.67 to 53.33   **8b.** 45.42 to 54.58   **8c.** 45.21 to 54.79   **8d.** 43.11 to 56.89   **9a.** z test   **9b.** 1.41   **9c.** retain H_0 ($z_{crit} = ±1.96$)   **10a.** 1.25   **10b.** z generally does, but in this case neither does

10c. Type II   **11a.** $t_{obt} = 2.12$ **11b.** retain H_0 ($t_{crit} = \pm 2.145$)   **11c.** .4973   **12a.** .895   **12b.** .8114   **12c.** yes

TRUE-FALSE QUESTIONS

T F   1. The t test for single samples is used when the population parameters μ and σ are unknown.

T F   2. In the t test s is used to estimate σ.

T F   3. In general t_{crit} is greater than z_{crit} at the same α level.

T F   4. The t test is more powerful than the z test.

T F   5. The t distribution is the same for all sample sizes.

T F   6. By definition the degrees of freedom for all statistical tests equals N - 1.

T F   7. The mean of the t distribution equals 0 for all sample sizes.

T F   8. When df = ∞ the t distribution is identical to the z distribution.

T F   9. The t test can only be applied to a nondirectional alternative hypothesis.

T F   10. For a two-tailed test, if $|t_{obt}| \geq |t_{crit}|$, reject H_0.

T F   11. The t test requires that the sampling distribution of X is normal.

T F   12. The confidence interval is one example of point estimation.

T F   13. The 95% confidence interval means that 95% of the time $\overline{X} = \mu$.

T F   14. The 99% confidence limit is always wider than the 95% confidence interval.

T F   15. If r_{obt} is not equal to zero, ρ cannot equal zero.

T F   16. For the t test of the significance of the correlation coefficient, the statistic of interest is $\overline{X}$.

T F   17. If r = -.10 and N = 8, ρ cannot equal +.10.

T F   18. If we conclude that there is a significant correlation in the population we may be making a Type I error.

T F 19. For the distribution of $\overline{X}$ to be normally distributed, the distribution of X must be normally distributed.

T F 20. The larger the sample size the more likely $\overline{X}$ is close to μ.

T F 21. At α = .05 there is more area in the critical region of rejection for the t distribution than for the z distribution.

T F 22. Changing sample size has no effect on power when using the t test.

T F 23. If ρ = 1.00, then r_{obt} must equal 1.00.

Answers: 1. F   **2.** T   **3.** T   **4.** F   **5.** F   **6.** F   **7.** T   **8.** T   **9.** F   **10.** T
11. F   **12.** F   **13.** F   **14.** T   **15.** F   **16.** F   **17.** F   **18.** T   **19.** F   **20.** T
21. F   **22.** F   **23.** T

SELF-QUIZ

1. The 95% confidence interval of the mean for $\overline{X}$ = 13.0, s = 1.6, and N = 21 is _____.

    a. 12.01 to 13.99
    b. 12.27 to 13.73
    c. 11.05 to 15.05
    d. 12.95 to 14.95

2. How many t distributions are there?

    a.   1
    b.   2
    c. 30
    d. one for each value of df

3. The mean of the t distribution equals _____.

    a. 0
    b. 1
    c. N
    d. N - 1

4. For a given value of alpha, the critical value of t is _____ than the critical value of z.

    a. no different
    b. less
    c. greater
    d. none of the above

5. As N gets _____, the critical value of t gets _____.

    a. larger; larger
    b. smaller; smaller
    c. larger; smaller
    d. smaller; larger

6. As N gets infinitely large, the critical value of t equals the critical value of z.

    a. true
    b. false

7. If $\mu = 30$, $\sigma = 5.2$, $\overline{X} = 28.0$, $s = 6.1$ and $N = 13$, the value of the most powerful statistic to test the significance of the sample mean is _____.

    a. -1.18
    b. 1.96
    c. 2.18
    d. -1.39

8. If the population parameters are known, the t test is _____ powerful than the z test.

    a. more
    b. less
    c. equally
    d. need more information

9. As N increases $s_{\overline{X}}$ becomes a _____ estimate of $\sigma_{\overline{X}}$.

    a. better
    b. worse
    c. meaningless
    d. biased

10. The proper use of the t test requires that _____.

    a. the sampling distribution of $\overline{X}$ is normal
    b. $N \geq 30$
    c. a and b
    d. a or b

11. The _____ confidence interval for the population mean is always wider than the _____ confidence interval for the population mean.

    a. 99%; 95%
    b. 95%; 99%

12. The sample mean is an example of _____.

    a. interval estimation
    b. point estimation
    c. confidence point
    d. confidence numbers

13. For N = 20, the degrees of freedom for the test of the significance of the correlation coefficient (r) is _____.

    a. 20
    b. 21
    c. 19
    d. 18

14. For N = 20, the value of r_{crit} for $\alpha = .05_{2\ tail}$ is _____.

    a. .4329
    b. .3783
    c. .4438
    d. .5614

15. If r = .84 and N = 5, the value of t_{obt} for the test of the significance of r is _____.

    a. 1.96
    b. 2.40
    c. 2.68
    d. 3.46

Answers: 1. b **2.** d **3.** a **4.** c **5.** c **6.** a **7.** d **8.** b **9.** a **10.** d
11. a **12.** b **13.** d **14.** c **15.** c

STUDENT'S t TEST FOR CORRELATED AND INDEPENDENT GROUPS

CHAPTER OUTLINE

I. The Two Condition Experiment

A. Types of two condition experiments.

   1. Correlated groups design. In the earlier chapters we analyzed this using the sign test.

   2. Independent groups design. We have analyzed this design using the Mann-Whitney U test.

B. Limitations of single sample design.

   1. At least one population parameter (μ) must be specified.

   2. Usually μ is not known.

   3. Even if μ were known, one cannot be certain that the conditions under which μ was calculated are the same for a new set of experimental conditions.

   4. These limitations are overcome in the two condition experiment.

II. Student's t Test for Correlated Groups

A. Requires repeated measures design (or correlated groups design).

     1. Each subject used for both conditions (e.g. before and after; control and experimental).

     2. Or pairs of subjects matched on one or more characteristics serve in both conditions.

B. Information used by t test.

     1. Information on magnitude of difference score.

     2. Information on direction of difference score.

C. Tests the assumption that the difference scores are a random sample from a population of difference scores having a mean of zero.

D. Similar to t test for single samples. The only change is that in this case we deal with difference scores instead of raw scores.

E. Equations:

$$t_{obt} = \frac{\overline{D}_{obt} - \mu_D}{s_D / \sqrt{N}}$$

$$t_{obt} = \frac{\overline{D}_{obt} - \mu_D}{\sqrt{\dfrac{SS_D}{N(N-1)}}}$$

where D = difference score (e.g. control score - experimental score)
           $\overline{D}$ = mean of the sample difference scores
           μ_D = mean of the population of difference scores (usually but not necessarily equal to 0)
           s_D = standard deviation of the sample difference scores
         SS_D = sum of squares of the sample difference scores
              = $\sum D^2 - (\sum D)^2 / N$
           N = number of difference scores

F. t test is more powerful than the sign test; therefore there is less chance of making a Type II error. Note: As a general rule one uses the most powerful statistical analysis appropriate to the data.

G. Assumptions for use requires that the sampling distribution of $\overline{D}$ is normally distributed. This can be achieved generally by:

1. $N \geq 30$, or
2. Population scores normally distributed

III. Independent Groups Design

A. Random sampling of subjects from population.

B. Random assignment to each condition.

C. No basis for pairing of scores.

D. Each subject tested only once.

E. Raw scores are analyzed.

F. Mann-Whitney U test analyzes separation between sample scores.

G. t test analyzes difference between sample means.

IV. Use of z Test for Independent Groups

A. Formula:

$$z_{obt} = \frac{(\overline{X}_1 - \overline{X}_2) - \mu_{\overline{X}_1 - \overline{X}_2}}{\sigma_{\overline{X}_1 - \overline{X}_2}}$$

where $\sigma_{\overline{X}_1 - \overline{X}_2} = \sqrt{\sigma_{\overline{X}_1}^2 + \sigma_{\overline{X}_2}^2} = \sqrt{\sigma^2 \left(\frac{1}{n_1} + \frac{1}{n_2} \right)}$

B. Assumptions.

1. Changing level of the independent variable is assumed to effect the mean of the distribution but not the standard deviation.

2. $\sigma_1^2 = \sigma_2^2 = \sigma^2$

C Characteristics of sampling distribution of the difference between sample means.

1. Assuming population from which samples are drawn is normal, then the distribution of the difference between sample means is normal.

2. $\mu_{\overline{X}_1 - \overline{X}_2} = \mu_1 - \mu_2$, where $\mu_{\overline{X}_1 - \overline{X}_2} =$ mean of the distribution of the difference between sample means.

3. $\sigma_{\overline{X}_1 - \overline{X}_2} = \sqrt{\sigma_{\overline{X}_1}^2 + \sigma_{\overline{X}_2}^2}$, where $\sigma_{\overline{X}_1 - \overline{X}_2} =$ the standard deviation of the difference between sample means. $\sigma_{\overline{X}_1}^2 =$ variance of the sampling distribution of the mean for samples of size n_1 taken from the first population. $\sigma_{\overline{X}_2}^2 =$ variance of the sampling distribution of the mean for samples of size n_2 taken from the second population.

D. To use z test one must know σ^2 which is rarely the case.

V. Student's t Test for Independent Groups

A. Used when σ^2 must be estimated. Uses a weighted average of the sample variances, s_1^2 and s_2^2, as the estimate with df as the weights.

B. General equation:

$$t_{obt} = \frac{(\overline{X}_1 - \overline{X}_2) - \mu_{\overline{X}_1 - \overline{X}_2}}{\sqrt{s_W^2 \left(\frac{1}{n_1} + \frac{1}{n_2}\right)}} = \frac{\overline{X}_1 - \overline{X}_2}{\sqrt{\left(\frac{SS_1 + SS_2}{n_1 + n_2 - 2}\right)\left(\frac{1}{n_1} + \frac{1}{n_2}\right)}}$$

where df = $n_1 + n_2 - 2 = N - 2$

C. Equation when $n_1 = n_2$:

$$t_{obt} = \frac{\overline{X}_1 - \overline{X}_2}{\sqrt{\frac{SS_1 + SS_2}{n(n-1)}}}$$

D. Assumptions for use of t test for independent groups.

    1. Sampling distribution of $\overline{X}_1 - \overline{X}_2$ is normally distributed, i.e. populations from which samples were taken must be normal.

    2. $\sigma_1{}^2 = \sigma_2{}^2$ **(homogeneity of variance)**.

E. Violations of assumptions. If $n_1 = n_2$ and $n \geq 30$, then the t test is robust if above assumptions are violated. If violations are extreme use Mann-Whitney U test.

VI. Power of t Test

A. The greater the effect of the independent variable, the higher the power.

B. Increasing sample size increases power.

C. High sample variability decreases power.

VII. Use of Correlated or Independent t

A. Correlated t advantageous when there is a high correlation between the paired scores.

B. Correlated t is advantageous when there is low variability in difference scores and high variability in raw .

C. Independent t is more efficient from df per measurement analysis.

D. Some experiments do not allow same subject to be used twice (i.e. comparing males vs. females).

CONCEPT REVIEW

In this chapter we have been studying methods of analyzing

data from the (1) _____ condition experiment. It has great   (1) two

advantage over the one condition experiment. We do not

need to specify any (2) _____ parameters to analyze the results.   (2) population

    In the repeated measures or (3) _____ (4) _____ design, a   (3) correlated
    (4) groups

(5) _____ score is calculated and analyzed. Though it is   (5) difference

common, it is not necessary that the (6) _____ subject be   (6) same

used under both conditions. We can (7) _____ the subjects on   (7) match

certain important characteristics instead. The (8) _____ test   (8) sign

utilizes information only on the (9) _____ of the difference   (9) direction

between scores; whereas the t test for correlated groups

utilizes information on both direction and (10) _____ of the   (10) magnitude

difference scores. In evaluating H_0, this t test usually tests the

assumption that the difference scores are a random (11) _____   (11) sample

from a population of difference scores having a mean of

(12) _____. The equation for the t test for correlated groups is: (12) zero

$$t_{obt} = \frac{(13)\,\underline{\quad} - \mu_D}{\dfrac{(14)\,\underline{\quad}}{\sqrt{(15)\,\underline{\quad}}}}$$

(13) $\overline{D}$
(14) s_D
(15) N

The computational formula is:

$$t_{obt} = \frac{(16)\,\underline{\quad} - \mu_D}{\sqrt{\dfrac{(17)\,\underline{\quad}}{(18)\,\underline{\quad}\,(N-1)}}}$$

(16) $\overline{D}$
(17) SS_D
(18) N

where (19) _____ = the difference score,   (19) D

    (20) _____ = the mean of the sample difference scores,   (20) $\overline{D}$

    (21) _____ = mean of the population of difference scores,   (21) μ_D

    (22) _____ = the standard deviation of the sample   (22) s_D
            difference scores,

(23) _____ = number of difference scores, (23) N

(24) _____ = D - $\overline{D}$ (24) d

For all t tests the decision rule for evaluating the null hypothesis

is the same:

If | (25) _____ **| ≥ | (26)** _____ **|, (27)** _____ **H₀** (25) t_{obt}
(26) t_{crit}
(27) reject

Compared to the sign test, the t test is (28) _____ powerful. (28) more

There is a (29) _____ probability of making a Type II error. (29) lower

As a general rule, one should use the most (30) _____ test (30) powerful

available to analyze the results of an experiment since this gives

the highest (31) _____ of rejecting H₀. To use the t test for (31) probability

correlated groups, the (32) _____ (33) _____ of $\overline{D}$ must be (32) sampling
(33) distribution

(34) _____ distributed. This means N should be ≥ (35) _____, (34) normally

or that the population scores themselves should be normally (35) 30

distributed.

The other major type of experimental design is the (36) _____ (36) independent

(37) _____ design. In this design there is no basis for (37) groups

(38) _____ scores. Each subject gets tested only (39) _____. (38) pairing
(39) once

With the z test we analyze the (40) _____ between the sample (40) difference

(41) _____. In order to use the z test to analyze the data (41) means

from an independent groups design, we would need to know the

value of the population parameter (42) _____. Since it is rarely (42) σ^2

known, we estimate it using the weighted (43) _____ of the (43) average sample variances. The weighting is done using the (44) _____

(45) _____ as the weights. The formula for the t test for

independent groups is:

$$t_{obt} = \frac{((46)\underline{\quad} - (47)\underline{\quad}) - \mu_{\overline{X}_1 - \overline{X}_2}}{\sqrt{(48)\underline{\quad}\left(\dfrac{1}{(49)\underline{\quad}} + \dfrac{1}{(50)\underline{\quad}}\right)}}$$

(46) $\overline{X}_1$
(47) $\overline{X}_2$
(48) s_W^2
(49) n_1
(50) n_2

Generally, we test the null hypothesis by assuming that

$\mu_{\overline{X}_1 - \overline{X}_2}$ equals (51) _____. Also, $s_W{}^2$ is the (52) _____

estimate of (53) _____. The general computational formula is:

(51) 0
(52) weighted

(53) σ^2

$$t_{obt} = \frac{(\overline{X}_1 - \overline{X}_2) - \mu_{\overline{X}_1 - \overline{X}_2}}{\sqrt{\left(\dfrac{(54)\underline{\ } + (55)\underline{\ }}{(56)\underline{\ } + (57)\underline{\ } - (58)\underline{\ }}\right)\left(\dfrac{1}{(59)\underline{\ }} + \dfrac{1}{(60)\underline{\ }}\right)}}$$

(54) SS_1
(55) SS_2
(56) n_1
(57) n_2
(58) 2
(59) n_1
(60) n_2

**where SS = (61) ____ - (62) ____/(63) ____ for each
    sample**

The df = N - (64) _____ where N = (65) _____ +

(66) _____.

(61) ΣX^2
(62) $(\Sigma X)^2$
(63) (N)

(64) 2
(65) n_1
(66) n_2

The computational formula when $n_1 = n_2$ is:

$$t_{obt} = \frac{(\overline{X}_1 - \overline{X}_2)}{\sqrt{\dfrac{S S_1 + S S_2}{(67)\underline{\ }\left((68)\underline{\ } - (69)\underline{\ }\right)}}}$$

(67) n
(68) n
(69) 1

If n_1 (70) _____ n_2, the more general equation must be used.   (70) $\neq$

To use the t test for independent groups we assume that the

sampling distribution of (71) _____ is normally distributed. We   (71) $\overline{X}_1 - \overline{X}_2$

also assume (72) _____ of variance. One way for this to occur   (72) homogeneity

is if the (73) _____ variable affects the means of the populations   (73) independent

but not their (74) _____ (75) _____. The t test is con-   (74) standard

                                                                                                                (75) deviations

sidered to be (76) _____, meaning it withstands violations of   (76) robust

the above assumptions under some conditions. The t test is

relatively (77) _____ to violations of the assumptions if   (77) insensitive

(78) _____ = (79) _____ and each sample $\geq$ (80) _____.   (78) n_1

                                                                                                                (79) n_2

In general with the t test, power can be affected by several   (80) 30

factors. The (81) _____ the effect of the independent variable,   (81) greater

the more likely the value of the (82) _____ of the t equation will   (82) numerator

be large. If other factors are constant this will increase the power.

Increasing the sample size will (83) _____ the denominator   (83) decrease

thereby (84) _____ power. As the sample variance increases   (84) increasing

the denominator of the t equation will (85) _____ which will   (85) increase

(86) _____ power.   (86) decrease

There are certain advantages and disadvantages to using the

t test for correlated or independent groups. When there is

(87) _____ variability in the (88) _____ scores, but   (87) high

                                                                                                                (88) raw

(89) _____ variability in the (90) _____ scores, it would   (89) low

                                                                                                                (90) difference

be preferable to use the t tests for correlated groups. If the

(91) _____ (92) _____ between the paired scores of con-   (91) correlation

dition 1 and 2 is high it would be advantageous to use the t test

for correlated groups. Many times it is not possible to use the

same subject in both conditons and often it is difficult to achieve

appropriate matching. Under these conditions we would use the

t test for (93) _____ (94) _____. The t test for independent

groups has as its major advantages the fact that it makes more

efficient use of (95) _____ of (96) _____ for each measure-

ment and its relative ease of recruiting (97) _____.

(92)   coefficient

(93)   independent
(94)   groups

(95)   degrees
(96)   freedom
(97)   subjects

EXERCISES

1. A dog food manufacturer believes that he may have found a way to improve the taste of the dog food made by his company. To test the new process, the company recruited 12 dogs. The dogs were given the original food and the time required to eat a standard portion of the food was measured. Two days later the same dogs were fed the same quantity of the new formula and the time was measured again. The following times were recorded:

Dog	Old Process (sec)	New Process (sec)
A	135	120
B	115	118
C	120	120
D	140	120
E	137	131
F	150	142
G	124	125
H	132	121
I	119	117
J	147	148
K	160	135
L	135	128

It is assumed that faster eating represents greater preference for the product.

   a. State the nondirectional alternative hypothesis.
   b. State the null hypothesis.
   c. What is the value of t_{obt}?
   d. What do you conclude using $\alpha = .05_{2\ tail}$?
   e. What is one obvious (okay, maybe it's not so obvious) defect with this study?

2. A political candidate wishes to determine if endorsing increased social spending is likely to affect her standing in the polls. She has access to data on the popularity of several other candidates who have endorsed increased spending. The data was available both before and after the candidates announced their positions on the issue. The data is as follows:

Candidate	Popularity Ratings	
	Before	**After**
1	42	43
2	41	45
3	50	56
4	52	54
5	58	65
6	32	29
7	39	46
8	42	48
9	48	47
10	47	53

Assuming no other factors were influencing the popularity ratings;

   a. State the nondirectional alternative hypothesis.
   b. State the null hypothesis.
   c. What is the value of t_{obt}?
   d. What might the candidate conclude using $\alpha = .01_{2\ tail}$?
   e. What type error might the candidate have made?

3. A medical researcher wishes to evaluate the effectiveness of a drug in prolonging the life of a red cell in a culture medium. Seven cultures are treated with saline (an inactive control substance) and seven others are treated with the new drug. The following survival times in days were recorded:

Control Group	Treatment Group
109	121
112	119
118	119
116	126
121	121
100	115
113	117

a. What kind of design is this?
b. State the directional alternative hypothesis.
c. State the null hypothesis.
d. What is the value of t_{obt}?
e. What do you conclude using $\alpha = .05_{1\ tail}$?

4. A worker in a neighborhood clinic wishes to assess the impact of showing an educational film on patient compliance in taking an antihypertension medication. The diastolic blood pressure (BP) is used as the dependent variable. During the study the medication dosage was kept constant. Blood pressure was measured one week prior to the film, then the film was shown, and the blood pressure measured again 3 weeks later. The data are shown below:

Patient	Diastolic BP (mmHg)	
	Before	After
1	110	100
2	105	95
3	98	88
4	100	92
5	89	83
6	82	86
7	113	100
8	102	101
9	101	96
10	118	112

a. State the nondirectional alternative hypothesis.
b. State the null hypothesis.
c. What is the value of t_{obt}?
d. What do you conclude using $\alpha = .01_{2\ tail}$?

5. Consider the following data collected from a repeated measures design study.

Before	After
5	17
8	25
5	19
7	32
6	26
9	8
10	9

a. Using $\alpha = .05_{2\ tail}$, what would you conclude about the null hypothesis using the t test?
b. What would you conclude using the sign test?
c. What does this indicate?

6. A biologist believes that temperature affects the croaking (noise making, not dying) behavior of frogs. A group of laboratory frogs are randomly divided into 2 groups and placed in identical terrariums. The control group of frogs' is kept at a constant 22°C. The experimental group is kept at a temperature of 30° C. The number of croaks emitted over a 10 minute period are counted. The data are indicated below:

Number of Croaks	
$22°$	$30°$
23	30
30	32
31	36
28	39
26	30
12	18
19	25
18	26

a. State the nondirectional alternative hypothesis.
b. State the null hypothesis.
c. What is the value of t_{obt}?
d. What do you conclude using $\alpha = .01_{2\ tail}$?
e. Can you think of another way to design this study to get more information for the biologist? Explain.

7. A teacher wishes to test the effects of using reinforcement versus a conventional method for maintaining classroom deportment. Twenty different teachers are instructed on the techniques to be used in the study; 10 on the conventional method

and 10 on using reinforcement. Data was gathered on the number of minutes of attentive behavior in an hour sample taken from a class of each teacher when the classes were being taught the same material. The classrooms were chosen because they were not significantly different in terms of socioeconomic status or past educational experiences. The data is as follows:

Attentive Behavior (Min)	
Conventional	Reinforcement
12	30
20	41
25	32
14	15
16	47
40	50
38	39
36	35
29	20
33	40

a. State the nondirectional alternative hypothesis.
b. State the null hypothesis.
c. What is the value of t_{obt}?
d. What do you conclude using $\alpha = .01_{2\ tail}$?

8. As an owner of a computer component factory you want to buy a new machine to make integrated circuits. You can buy one of two types of machines. You obviously want to buy the better machine (we'll assume they are comparable in price and reliability). You rent both machines for 8 days and make sample runs on both machines. The following data are the number of defective components per 1000 on each day.

Machine A	Machine B
2	4
1	3
3	5
1	3
2	4
4	6
6	8
0	2

a. What is the value of t_{obt}?

   b. Which machine would you buy? Use $\alpha = .05_{2 \text{ tail}}$. Assume independence between each day's run of each machine.

   c. Looking at this data, what would you do now?

9. A drug company wants to evaluate the effects of a new tonic on stimulating new hair growth. Two comparable groups of male volunteers are tested. Group 1 is given a mixture of chicken fat and bear grease as a control tonic. Group 2 gets the new tonic. The number of new hairs are counted one month after the onset of tonic use. The results are shown below:

New Tonic	Control Tonic
6	5
12	13
10	6
14	8
13	25
8	50
65	

   a. State the directional alternative hypothesis.

   b. State the null hypothesis.

   c. What is the value of t_{obt}?

   d. What do you conclude? $\alpha = .05_{1 \text{ tail}}$.

   e. What type error might you be making?

10. An automotive engineer wishes to determine the fuel efficiency of a test engine compared to the standard production model. The following data (miles/gallon) were obtained under standard conditions. Assume independence between scores in each group.

	Engine	
Standard		Test
25		26
24		27
19		20
21		22
22		23
26		27
25		26
23		

a. Test the null hypothesis that there is no difference in the fuel efficiencies of these two engines. What do you conclude using $\alpha = .05_{2\ tail}$?

b. What type of error might you be making?

Answers: 1a. The change in manufacturing process affects the time it takes dogs to eat a meal. $\mu_D \neq 0$ **1b.** The change in manufacturing process has no effect on the time dogs take to eat a meal. $\mu_D = 0$ **1c.** $t_{obt} = 2.88$ **1d.** $t_{crit} = \pm2.201$; therefore reject H_0 **1e.** The order of food presentation is always the same. It would be better if half the dogs were given the new process first so if conditions in the lab are different on different days, it won t have a systematic effect on the results.

2a. Endorsing increased social spending affects popularity ratings. $\mu_D \neq 0$ **2b.** Endorsing increased social spending has no effect on popularity ratings. $\mu_D = 0$ **2c.** -3.10   **2d.** $t_{crit} = \pm3.250$; retain H_0 **2e.** Type II

3a. Independent groups design   **3b.** The new drug prolongs the life of red cells in a culture medium. $\mu_1 > \mu_2$ **3c.** The new drug does not prolong the life of red cells in a culture medium $\mu_1 \leq u_2$ **3d.** $t_{obt} = -2.40$ **3e.** $t_{crit} = -1.782$; reject H_0

4a. The film affects patient compliance. $\mu_D \neq 0$ **4b** The film has no effect on patient compliance as measured by diastolic blood pressure; any change was due to chance alone. $\mu_D = 0$.   **4c.** $t_{obt} = 4.12$ **4d.** $t_{crit} = \pm3.250$; reject H_0

5a. $t_{obt} = -3.25$, $t_{crit} = \pm2.447$; reject H_0 **5b.** $.4532 > .05$; retain H_0 **5c.** t test for correlated measures has more power.

6a. Temperature affects the croaking behavior of laboratory frogs.   **6b.** Temperature has no effect on the croaking behavior of laboratory frogs. $\mu_1 = \mu_2$ **6c.** $t_{obt} = -1.85$ **6d.** $t_{crit} = \pm2.977$; retain H_0 **6e.** Use a regression experiment. Take the same frogs and measure the number of croaks at several different temperatures. Construct the regression line through the points (X = temp., Y = # of croaks). One could also calculate the correlation coefficient and test the significance of r.

7a. There is a difference in the attentive behavior of the two classes.   **7b.** There is no difference in the attentive behavior of the two classes. $\mu_1 = \mu_2$. NOTE: The experiment is making the supposition that any difference is due to the use of either reinforcement or conventional techniques and not any other reason. This may not be a good assumption. **7c.** $t_{obt} = -1.79$ **7d.** $t_{crit} = \pm2.878$; retain H_0

8a. $t_{obt} = 2.08$; $t_{crit} = \pm2.145$ **8b.** At this point there is no significant difference apparent between the two machines   **8c.** One might well continue testing the two machines. It looks like the sample may be too small to detect any differences (i.e., experiment not powerful enough.) One or two more points might have allowed us to reject H_0.

9a. The new hair tonic increases new hair growth. $\mu_1 > \mu_2$ **9b.** The new tonic does not increase new hair growth. $\mu_1 \leq \mu_2$ **9c.** $t_{obt} = -.04$ **9d.** $t_{crit} = 1.796$; retain H_0 **9e.** Type II.

10a. $t_{obt} = -0.99$: $t_{crit} = 2.160$. Retain H_0.   **10b.** Type II.

TRUE-FALSE QUESTIONS

T F   1. In a repeated measures design the difference scores are used in the analysis, not the raw scores.

T F   2. If H_0 is true, then μ_D must equal 0 for a correlated groups design.

T F   3. H_1 cannot be true if $\overline{D} = 0$ for a correlated groups design.

T F   4. $SS_D = \Sigma (D - \overline{D})^2$

T F   5. The t test for correlated groups is generally more powerful than the sign test.

T F   6. When several tests are appropriate for analyzing data, it is best to use the least powerful test so as to minimize the probability of making a Type I error.

T F   7. The t test for correlated groups requires the sampling distribution of $\overline{X}$ be normally distributed.

T F   8. In the independent groups design there is no matching of subjects and each subject is tested only once.

T F   9. With both the t test for independent groups and the Mann-Whitney U test, the difference between the sample means is tested.

T F   10. One way of stating H_0 for the t test for independent groups is $\mu_1 - \mu_2 = 0$. (Assume H_1 is nondirectional)

T F   11. In the independent groups t test, $s_{\overline{X}_1 - \overline{X}_2}$ is an estimate of $\sigma_{\overline{X}_1 - \overline{X}_2}$

T F   12. In the independent groups t test an average of s_1^2 and s_2^2, weighted by the degrees of freedom, is used to estimate σ^2.

T F   13. For both the t test for repeated measures and independent groups, df $= n_1 + n_2 - 2$.

T F 14. If $n_1 = n_2$ and σ^2 is unknown, then the z test can be used to analyze the data from an independent groups design.

T F 15. The t test for independent groups assumes homogeneity of variance.

T F 16. The t test assumes that the independent variable affects either the mean of the dependent variable or the standard deviation, but not both.

T F 17. If a violation of the homogeneity of variance assumption occurs, it is always necessary to use the Mann-Whitney U test for an independent groups design.

T F 18. Other factors held constant, for an independent groups design, the larger the value of $\overline{X}_1 - \overline{X}_2$, the greater the likelihood of rejecting H_0.

T F 19. For the independent groups design, increasing sample variance decreases power.

T F 20. Generally, if r between the scores of both groups is high then the t test for correlated samples is a powerful statistical test.

T F 21. If r between the scores of both groups is high and negative then the t test for correlated samples is not a powerful statistical test.

T F 22. The higher df are, the lower t_{crit} becomes for rejecting H_0.

T F 23. There is a greater chance of making a Type I error using $\alpha = .05$ when applying the t test for independent groups then when applying the Mann-Whitney U test.

T F 24. It is not necessary to measure the population parameters to apply the t test for either the independent or correlated groups designs.

T F 25. If $\overline{X}_1 - \overline{X}_2 = 0$, then H_0 must be true.

Answers: 1. T **2.** F **3.** F **4.** T **5.** T **6.** F **7.** F **8.** T **9.** F **10.** T **11.** T **12.** T **13.** F **14.** F **15.** T **16.** F **17.** F **18.** T **19.** T **20.** T **21.** F **22.** T **23.** F **24.** .T **25.** F.

SELF-QUIZ

1. A major advantage to using a two condition experiment (e.g. control and experimental groups) is _____.

   a. the test has more power
   b. the data are easier to analyze
   c. the experiment does not need to know population parameters
   d. a and b

2. In testing the null hypothesis, the correlated t test allows one to utilize information on _____ in the test of significance.

   a. magnitude and direction
   b. magnitude only
   c. direction only
   d. separation between the groups

3. In a correlated t test, if the independent variable has no effect, the sample difference scores are a random sample from a population where the mean difference score (μ_D) equals _____.

   a. 0
   b. 1
   c. N
   d. cannot be determined

4. In analyzing the results of the correlated t test, the _____ scores are analyzed.

   a. standardized
   b. normalized
   c. raw
   d. difference

5. The t test is _____ likely to result in a Type II error than the sign test for a repeated measures design.

   a. more
   b. less
   c. equally
   d. probably

6. Which of the following tests analyzes the difference between the means of two independent samples?

    a. correlated t test
    b. t test for independent groups
    c. Mann-Whitney U test
    d. all of the above

7. In an independent groups design the nondirectional alternative hypothesis states
    _____.

    a. $\mu_1 > \mu_2$
    b. $\mu_1 < \mu_2$
    c. $\mu_1 \neq \mu_2$
    d. $\mu_1 = \mu_2$

8. The z test for independent groups is almost never used because it requires that
    _____ be known.

    a. μ_1
    b. $\overline{X}_1$
    c. σ^2
    d. s^2

9. In the t test for independent samples, there are _____ degrees of freedom.

    a. $n_1 - 1$
    b. $n_1 + n_2$
    c. $n_1 - n_2 + 2$
    d. $n_1 + n_2 - 2$

10. The t test assumes that the independent variable affects the _____ of the populations.

    a. means
    b. standard deviations
    c. a and b
    d. none of the above

11. The use of the t test for independent groups assumes _____.

    a. $\overline{X}_1 - \overline{X}_2$ is normally distributed
    b. $\sigma_1^2 = \sigma_2^2$
    c. $\overline{D}$ is normally distributed
    d. all of the above
    e. a and b

12. The power of a correlated t test increases if the correlation between the paired scores is _____.

    a. 0
    b. high
    c. low
    d. cannot be determined

13. The sampling distribution of $\overline{X}_1 - \overline{X}_2$ has a mean value equal to _____.

    a. 0
    b. $\mu_1 - \mu_2$
    c. N
    d. $s_{\overline{X}_1 - \overline{X}_2}$

14. If $n_1 = n_2$ and n is relatively large, then the t test is relatively robust against _____.

    a. violations of the assumptions of homogeneity of variance and normality
    b. violations of random samples
    c. traffic violations
    d. violations by the forces of evil

15. Five students were tested before and after taking a class to improve their study habits. They were given articles to read which contained a known number of facts in each story. After the story each student listed as many facts as he/she could recall. The following data was recorded.

Before	10	12	14	16	12
After	15	14	17	17	20

    What do you conclude using $\alpha = .05_{2 \text{ tail}}$?

    a. reject H_0
    b. retain H_0

16. A physical therapist wants to know if football players (guards in this experiment) recover full strength following injuries. A group of injury free guards were given a strength test as were a group of guards who had finished a rehabilitation program following an injury. The groups were matched in height and weight. The following strength ratings were recorded.

Non-injured	Injured
302	297
306	300
258	250
239	245
280	270
265	262
274	

What do you conclude using $\alpha = .01_{2\ tail}$?

a. reject H_0
b. retain H_0

Answers: 1. c **2.** a **3.** a **4.** d **5.** b **6.** b **7.** c **8.** c **9.** d **10.** a **11.** e **12.** b **13.** b **14.** a **15.** a (t_{obt}= -3.062; $t_{crit} = \pm2.776$) **16.** b ($t_{obt} =$ 0.319; $t_{crit} = \pm3.106$).

16

INTRODUCTION TO THE ANALYSIS OF VARIANCE

CHAPTER OUTLINE

I. Introduction - F Distribution

A. F is a ratio of two independent variance estimates.

B. Sampling distribution of F.

1. Take all possible values of size n_1 and n_2 from population.

2. Estimate population σ^2 from each of the samples using s_1^2 and s_2^2.

3. Calculate F_{obt} for all possible combinations of s_1^2 and s_2^2.

4. Calculate $p(F)$ for each different value of F_{obt}.

C. Since F is a ratio of variance estimates, it will never be negative (i.e. squared numbers are always positive).

D. F distribution is positively skewed.

E. With equal n's, median F value equals one.

F. F distribution is a family of curves for each combination of df.

G. Degrees of freedom for numerator = $n_1 - 1$; for denominator, $df = n_2 - 1$.

II. ANOVA

A. Used to analyze data from experiments which use more than two groups or conditions (can also be used for analyzing two conditions).

B. Used instead of many pairwise t tests in order to hold the probability of making a Type I error at α.

III. Overview of One-Way ANOVA Technique

A. F test allows us to make one overall comparison which tells whether there is a significant difference between the means of the groups.

B. ANOVA can be used for:

1. Independent groups design also called simple randomized group design or the one-way analysis of variance -- independent groups design or single factor experiment, independent groups design.

2. Repeated measures design.

C. In the independent groups design, there are as many groups as there are levels of the independent variable.

D. Hypothesis testing:

1. H_1 is nondirectional.

2. H_0 states that the different conditions are all equally effective, i.e., $\mu_1 = \mu_2 = \mu_3 = \cdots = \mu_k$

E. ANOVA assumes that the independent variable affects only the mean of the scores but not the variance, i.e.,

$$\sigma_1^2 = \sigma_2^2 = \sigma_3^2 = \cdots = \sigma_k^2$$

F. ANOVA partitions total variability of data (SS_T) into the variability that exists within each group (SS_W) and the variability between groups (SS_B). The SS stands for sum of squares.

G. SS_B and SS_W are both used as independent estimates of the H_0 population variance.

H. F ratio:

$$F_{obt} = \frac{\text{between-groups variance extimate } (s_B^2)}{\text{within-groups variance estimate } (s_W^2)}$$

I. s_B^2 increases with magnitude of effect of independent variable while s_W^2 is unaffected. The larger the F ratio the more unreasonable H_0 becomes.

J. **If $F_{obt} \geq F_{crit}$, reject H_0.**

IV. **Within-Groups Variance Estimate, s_W^2**

A. Analagous to s_W^2 for t test but is for ≥ 2 groups. It is an estimate of σ^2.

B. Conceptual equation:

$$s_W^2 = \frac{S S_1 + S S_2 + S S_3 + \cdots + S S_k}{N - k}$$

where $N = n_1 + n_2 + n_3 + \cdots + n_k$ and $N - k = df_W$

C. SS_W is the within-groups sum of squares and is the numerator of s_W^2.

$$s_W^2 = SS_W/df_W$$

V. **Between-Groups Variance Estimate, s_B^2**

A. If H_0 is true, the variance between the means of the samples is an estimate of $\sigma 2$.

B. Using $s_{\bar{X}}^2$ as estimate for $\sigma_{\bar{X}}^2$, we can get second estimate of σ^2.

C. Conceptual equation:

$$s_B^2 = \frac{n\Sigma(\overline{X}-\overline{X}_G)^2}{k-1} = \frac{n\left[(\overline{X}_1-\overline{X}_G)^2 + (\overline{X}_2-\overline{X}_G)^2 + (\overline{X}_3-\overline{X}_G)^2 + \cdots + (\overline{X}_k-\overline{X}_G)^2\right]}{k-1}$$

where $\overline{X}_G$ is the overall mean of all the scores.

D. Numerator of s_B^2 is called the between-groups sum of squares SS_B. The denominator is the df for s_B^2:

$$s_B^2 = SS_B/df_B$$

E. As magnitude of effect of the independent variable increases, $\Sigma(\overline{X}-\overline{X}_G)^2$ increases, thereby increasing SS_B and s_B^2.

VI. F Ratio

A. s_B^2 increases with magnitude of effect of independent variable.

B. Since independent variable effects only the mean not the variance, s_W^2 does not change.

C. Since $F = s_B^2/s_W^2$, as the independent variable effect increases, the F ratio increases.

D. If H_0 is true, F is expected to equal 1. The larger F becomes, the more reasonable it is that the independent variable has an effect.

E **F = (σ^2 + effects of independent variable)/σ^2.**

F. If $F < 1$, retain H_0.

G. **F = t^2**. ANOVA can be used when $k = 2$ instead of t test.

VII.   Calculating F_{obt}

A. Calculate SS_B:

$$SS_B = \left[\frac{(\Sigma X_1)^2}{n_1} + \frac{(\Sigma X_2)^2}{n_2} + \frac{(\Sigma X_3)^2}{n_3} + \cdots + \frac{(\Sigma X_k)^2}{n_k}\right] - \frac{\overset{\text{all scores}}{(\Sigma X)^2}}{N}$$

B. Calculate SS_W:

$$SS_W = \overset{\substack{\text{all} \\ \text{scores}}}{\sum} X^2 - \left[\frac{(\sum X_1)^2}{n_1} + \frac{(\sum X_2)^2}{n_2} + \frac{(\sum X_3)^2}{n_3} + \cdots + \frac{(\sum X_k)^2}{n_k} \right]$$

C. Calculate SS_T as a check for work (remember $SS_T = SS_B + SS_W$):

$$SS_T = \overset{\substack{\text{all} \\ \text{scores}}}{\sum} X^2 - \frac{\left(\overset{\substack{\text{all} \\ \text{scores}}}{\sum} X \right)^2}{N}$$

D. Calculate df:

$$df_B = k - 1$$

$$df_W = N - k$$

$$df_T = N - 1$$

E. Calculate s_B^2:

$$s_B^2 = SS_B / df_B$$

F. Calculate s_W^2:

$$s_W^2 = SS_W / df_W$$

G. Calculate F_{obt}:

$$F_{obt} = s_B^2 / s_W^2, \text{ then evaluate with } F_{crit}$$

VIII. Assumptions Underlying Use of ANOVA

A. Populations from which samples are drawn are normally distributed.

B. Samples have equal variances (homogeneity of variance).

C. Violations of assumptions:

   1. F is robust if sample sizes are equal.

   2. F is minimally affected by violations of population normality.

IX. Two-Way ANOVA

A. A more sophisticated technique used to investigate more than one factor (or independent variable). A factorial experiment is one in which the effect of two or more factors are assessed in one experiment. In a factorial experiment the treatments used are combinations of the levels of both factors.

B. Allows us to evaluate the effect of two independent variables (e.g. A and B) and the interaction between them (A x B).

C. <u>Main effects:</u> The effect of Factor A averaged over the levels of Factor B and the effect of Factor B averaged over the levels of Factor A are called main effects.

D. Interaction effects occur when the effect of one factor is not the same at all levels of the other factor.

E. Analyzing data:

   1. Calculate four variance estimates:

      a. s_W^2; the within-groups variance estimate

      b. s_R^2; row variance estimate (sensitive to A effects)

      c. s_C^2; column variance estimate (sensitive to B effects)

      d. s_{RC}^2 ; row by column variance estimate (sensitive to interaction effects)

   2. Test for effects:

$$\text{For A:}\ \ F_{obt} = s_R^2 / s_W^2$$

$$\text{For B:}\ \ F_{obt} = s_C^2 / s_W^2$$

For interaction between A and B: $F_{obt} = s_{RC}^2/s_W^2$

CONCEPT REVIEW

Instead of using the mean for hypothesis testing, the F test

uses the (1) _____. The F statistic is the (2) _____ of two

independent variance estimates of the same population variance.

The F test has a sampling distribution which gives all possible

values of (3) _____ along with the (4) _____ for each

value, assuming sampling is (5) _____ from the population.

The F test has (6) _____ values for degrees of freedom, one

for each variance estimate. The F ratio is always a (7) _____

number. When n's are (8) _____ the (9) _____ value of F will

equal 1. Like the t test, there are a (10) _____ of curves for F

depending on the different combinations of df_1 and df_2.

The analysis of variance, abbreviated (11) _____ is appro-

priate to analyze data from experiments which use more than

(12) _____ conditions or levels of the (13) _____ variable.

In the ANOVA the independent variable is often referred to as

a (14) _____. The ANOVA is used to analyze results from

an experiment instead of a series of t tests to avoid (15) _____

the probability of making a Type (16) _____ error. The use of

the ANOVA (F test) allows us to make (17) _____ overall

comparison between the (18) _____ of the groups. (Note: We

(1) variance
(2) ratio

(3) F
(4) probability
(5) random

(6) two

(7) positive

(8) equal
(9) median
(10) family

(11) ANOVA

(12) two
(13) independent

(14) factor

(15) increasing

(16) I

(17) one

(18) means

are testing the (19) _____, not the (20) _____ of the groups.)

The simplest form of ANOVA is called the (21) _____-

(22) _____ ANOVA. It is used in an (23) _____ groups

design. Though it is not required it is preferable to have equal

(24) _____ in each group. The alternative hypothesis in the

ANOVA is (25) _____.

The one-way ANOVA (26) _____ the total variability, de-

signated (27) _____ into two sources; the variability that exists

within each group, called the (28) _____-(29) _____

(30) _____ of (31) _____, SS_W and the (32) _____

that exists between the groups, called the (33) _____ -groups

(34) _____ of (35) _____, SS_B. Each sum of squares

is used for an independent (36) _____ of the H_0 population

(37) _____. The estimate based on the within-groups varia-

bility is called the within-groups variance estimate, (38) _____.

The estimate based on the between-groups variability is called

the between-groups variance estimate, (39) _____. The F

ratio, conceptually is:

$$F_{obt} = (40) \text{_____} / (41) \text{_____}$$

The (42) _____ -groups variance (43) _____ with the mag-

nitude of the effect of the independent variable while the

(44) _____-groups variance is unaffected. Therefore,

(19)	means
(20)	variances
(21)	one
(22)	way
(23)	independent
(24)	n's
(25)	nondirectional
(26)	partitions
(27)	SS_T
(28)	within
(29)	groups
(30)	sum
(31)	squares
(32)	variability
(33)	between
(34)	sum
(35)	squares
(36)	estimate
(37)	variance
(38)	s_W^2
(39)	s_B^2
(40)	s_B^2
(41)	s_W^2
(42)	between
(43)	increases
(44)	within

the (45) _____ the F ratio the more (46) _____ the null hypo-   (45) larger
(46) unreasonable

thesis becomes. To evaluate F_{obt} we apply the same decision

rules as usual, namely:

If (47) _____ ≥ (48) _____, reject H_0   (47) F_{obt}
(48) F_{crit}

In this case we don't have to worry about a (49) _____ value of   (49) negative

F. If the value of F is (50) _____ 1, we do not have to bother   (50) less than

to evaluate it since we know the probability will be (51) _____   (51) greater

than α.

The within-groups variance estimate is determined the same

way as for the t test except we allow for more groups.

Conceptually:

$$s_W^2 = \frac{S S_1 + S S_2 + \cdots + (52) \underline{\quad}}{(n_1 - 1) + (n_2 - 1) + \cdots + (n_k - 1)}$$ (52) SS_k

where (53) _____ = the number of groups. The numerator of   (53) k

s_W^2 above is called the within-groups (54) _____ of   (54) sum

(55) _____. For s_W^2 there are (56) _____ df. The   (55) squares
(56) N - k

computational formula for SS_W is:

$$S S_W = (57) \underline{\quad} - \left[\frac{(58) \underline{\quad}}{n_1} + \frac{(59) \underline{\quad}}{n_2} + \frac{(60) \underline{\quad}}{n_3} + \cdots + \frac{(61) \underline{\quad}}{n_k} \right]$$

(57) $\sum\limits^{all\ scores} X^2$
(58) $(\sum X_1)^2$
(59) $(\sum X_2)^2$
(60) $(\sum X_3)^2$
(61) $(\sum X_k)^2$

We can use SS_W in a simple formula for $s_W{}^2$:

$$s_W{}^2 = \text{(62)} \underline{\hspace{2cm}} / \text{(63)} \underline{\hspace{2cm}}$$

(62) SS_W
(63) $N - k$ or df_W

The between-groups variance conceptual formula looks much like the formula for the sample variance:

$$s_B{}^2 = \frac{\text{(64)}\underline{\hspace{0.5cm}} \sum \left(\text{(65)}\underline{\hspace{0.5cm}} - \text{(66)}\underline{\hspace{0.5cm}} \right)^2}{\text{(67)}\underline{\hspace{0.5cm}} - 1}$$

(64) n
(65) $\bar{X}$
(66) $\bar{X}_G$
(67) k

Expanding this equation gives the following conceptual formula:

(68) $\bar{X}_G$
(69) $\bar{X}_2$
(70) $\bar{X}_G$

$$s_B{}^2 = \frac{n\left[\left(\bar{X}_1 - \text{(68)}\underline{\hspace{0.5cm}}\right)^2 + \left(\text{(69)}\underline{\hspace{0.5cm}} - \text{(70)}\underline{\hspace{0.5cm}}\right)^2 + \cdots + \left(\text{(71)}\underline{\hspace{0.5cm}} - \text{(72)}\underline{\hspace{0.5cm}}\right)^2 \right]}{\text{(73)}\underline{\hspace{0.5cm}} - 1}$$

(71) $\bar{X}_k$
(72) $\bar{X}_G$
(73) k

with $k - 1$ (74) _____ of (75) _____. The numerator of the above equation is (76) _____. $\bar{X}_G$ is the (77) _____ mean. One should now be able to see how $s_B{}^2$ (more specifically SS_B) (78) _____ as the effect of the independent variable (79) _____.

(74) degrees
(75) freedom
(76) SS_B
(77) grand or overall

(78) increases

(79) increases

One of the assumptions of the analysis of variance is that the independent variable affects only the (80) _____, not the (81) _____, of each group. (82) _____ does not change as

(80) mean

(81) variance
(82) $s_W{}^2$ or SS_W

a result of the independent variable. s_W^2 and s_B^2 are both

independent estimates of (83) _____. Therefore:

(83) σ^2

$$\mathbf{F_{obt}} = \mathbf{s_B}^2/\mathbf{s_W}^2 = (\sigma^2 + (84) \text{_____})/\sigma^2$$

(84) effects of the independent variable

The larger the value of F_{obt} the (85) _____reasonable to

assume H_0 is false.

(85) more

We stated earlier that the total variability or total sum of squares

can be partitioned into (86) _____ parts in the one-way ANOVA.

Therefore,

(86) two

$$\mathbf{SS_T} = (87) \text{ _____} + (88) \text{ _____}$$

(87) SS_W
(88) SS_B

Except to check our work (a very good idea) we only need to

calculate two of the above and we can easily solve for the third

quantity. We can calculate SS_T as follows:

$$S\,S_T = \overset{\substack{\text{all} \\ \text{scores}}}{\sum} (89) \text{\_\_\_} - \frac{\left(\overset{\substack{\text{all} \\ \text{scores}}}{\sum} (90) \text{\_\_\_}\right)^2}{(91) \text{\_\_\_}}$$

(89) X^2
(90) X
(91) N

For example, if $SS_T = 1920$ and $SS_B = 1805$, SS_W would equal

(92) _____. The total degrees of freedom equals

(92) 115

(93) _____.

(93) $N - 1$

A slightly different way to look at the partitioning of the var-

iability involves the deviation of individual score from the

(94) _____ mean $(\overline{X} - \overline{X}_G)$. This deviation from the grand mean

(94) grand

can be thought of as the deviation of the score from its own

(95) _____ mean $(\overline{X} - \overline{X}_k)$, plus the deviation of that group from   (95) group

the (96) _____ mean $(\overline{X}_k - \overline{X}_G)$. The first deviation $(\overline{X} - \overline{X}_k)$ con-   (96) grand

tributes to (97) _____ and the second $(\overline{X}_k - \overline{X}_G)$ contributes   (97) SS_W

to (98) _____.   (98) SS_B

As with the t test there are certain assumptions to be consider-

ed. First, in using the ANOVA one assumes the populations

from which the samples are drawn are (99) _____ distributed   (99) normally

Second, one assumes the samples are drawn from populations

of (100) _____ variances. Like the t test, the ANOVA is   (100) equal

(101) _____. It is minimally affected by violations of   (101) robust

(102) _____ (103) _____. It is also relatively insensitive to   (102) population
(103) normality

violations of (104) _____ of variance if sample sizes in   (104) homogeneity

the groups are (105) _____.   (105) equal

In the special case where k = (106) _____, one can use   (106) 2

either the ANOVA or the (107) _____ test. In this case, the   (107) t

following relationship will hold:

$$\mathbf{F_{obt}} = (108) \ _____$$ (108) t_{obt}^2

Two-Way ANOVA

A slightly more complicated way of analyzing experiments is the

two-way ANOVA. The one-way ANOVA examines one independent

variable or (109) _____ with several (110) _____ of that   (109) factor
(110) levels

independent variable. The two-way ANOVA allows us to

evaluate in (111) _____ experiment the effect of

(112) _____ independent variables and the (113) _____

between them. Such experiments are called (114) _____

experiments. In a factorial experiment the treatments used are

(115) _____ of the levels of both factors. In such a design the

effect of Factor A averaged over the levels of Factor B and the

effect of Factor B averaged over the levels of Factor A are called

(116) _____ (117) _____. An (118) _____ effect occurs

when the effect of one factor is not the (119) _____ at all

levels of the other factor.

In analyzing data from a two-way ANOVA, we determine

(120) _____ variance estimates. They are (121) _____,

(122) _____, (123) _____ and (124) _____. The

estimate s_W^2 is the within-(125) _____ variance estimate and

is similar to s_W^2 in the simple one-way ANOVA. It becomes the

standard against which the other estimates are (126) _____.

s_R^2 is called the (127) _____ variance estimate and it is

sensitive to the effects of variable (128) _____. s_C^2 is the

(129) _____ variance estimate and it is sensitive to the effects

of variable (130) _____. These are both (131) _____

effects. The estimate (132) _____ is the (133) _____ by

(134) _____ variance estimate. It is sensitive to the

(135) _____ effects of variables A and B. The following

(111) one

(112) two
(113) interaction
(114) factorial

(115) combinations

(116) main
(117) effects
(118) interaction
(119) same

(120) four
(121) s_W^2
(122) s_R^2
(123) s_C^2
(124) s_{RC}^2
(125) cell

(126) compared

(127) row

(128) A

(129) column

(130) B
(131) main
(132) s_{RC}^2
(133) row
(134) column

(135) interaction

tests can be made.

For variable A:

$$F_{obt} = (136) \text{_____} / s_W^2$$ (136) s_R^2

For variable B:

$$F_{obt} = (137) \text{_____} / s_W^2$$ (137) s_C^2

For the interaction between A and B:

$$F_{obt} = (138) \text{_____} / s_W^2$$ (138) s_{RC}^2

The values of F_{obt} are evaluated against appropriate values of

(139) _____. The main advantage of the two-way ANOVA is   (139) F_{crit}

that we can do (140) _____ one-way experiments plus   (140) two

we are able to evaluate the (141) _____ between two   (141) interaction

(142) _____ variables.   (142) independent

EXERCISES

1. Why is SS_W insensitive to the effects of the independent variable?

2. Assuming that the sample sizes are equal, N = 12 and that k = 3, consider the following data:

$$s_1 = 8.113, \quad s_2 = 8.816, \quad s_3 = 7.258$$

   a. What is the value of s_W^2
   b. What is the value of SS_W?

3. If $SS_T = 106$ and $SS_W = 24$, what is the value of SS_B?

4. A researcher is interested in knowing if different types of treatments have an effect on the number of relapses for clients with the diagnosis of alcoholism. He randomly selects a group of 24 clients from a waiting list at a local mental health center who consent to the study. Then he randomly assigns 6 subjects to each treatment group. Group 1 is a waiting list control group. They are given no treatment. Group 2 receives a medication which causes illness when alcohol is consumed. Group 3 gets cognitive behavior therapy, and Group 4 gets cognitive behavior therapy plus the drug. The researcher then records the number of relapses in the next 12 months. The following data are obtained.

Group 1	Group2	Group 3	Group 4
12	4	7	1
14	6	5	3
10	6	5	0
8	6	2	2
9	3	1	2
10	2	1	1

a. State the alternative hypothesis.
b. State the null hypothesis.
c. What is the value of SS_W?
d. What is the value of SS_B?
e. What is the value of SS_T?
f. What is the value of F_{obt}?
g. What do you conclude, using $\alpha = .05$?

5. Is it possible to have a significant result with $k = 2$ using the t test, but when reanalyzing the data using the F test obtain a nonsignificant result? (Assume α is the same, the design is the same and you don't make a calculation error.)

6. The following data were obtained from an experiment to test the effect of different levels of variable A on heart rate (beats per minute).

	Levels of A	
Group 1	Group 2	Group 3
83	79	90
82	86	80
82	72	70
85	73	65
87	70	79

a. What is the value of s_B^2?

b. What is the value of F_{crit}, $\alpha = .01$?

c. What is the value of F_{obt}?

d. What do you conclude using $\alpha = .01$?

7. A dentist wants to know if the flavor of toothpaste affects how often children brush their teeth, and therefore the number of cavities. She makes up 3 flavors of toothpaste, cherry, bubblegum and spinach and observes the following numbers of cavities at a 6-month check-up:

Cherry	Bubblegum	Spinach
7	5	8
2	2	9
8	6	6
5	4	2
3	1	9
3	8	5
0	4	3

a. State H_1.

b. State H_0.

c. What is the value of s_B^2?

d. What is the value of s_W^2?

e. What is the value of F_{crit}, $\alpha = .05$?

f. What is the value of F_{obt}?

g. What do you conclude using $\alpha = .05$?

8. Complete the following table where indicated?

Source	S S	df	s^2	F
Between	117.17	(b)__	(c)_____	(e)_____
Within	(a)_____	9	(d)_____	
Total	177.67	11		

9. Analyze the following data from an independent groups design using ANOVA.

Control	Experimental
50	45
60	38
51	40
52	48
51	50
49	40
50	44

a. What is the value of df_W?
b. What is the value of df_B?
c. What is the value of F_{obt}?
d. What would the value of t_{obt} be if one analyzed the data using the t test? Do a t test on the above data to confirm your results.
e. Would you reject H_0 using $\alpha = .05$?

10 Given the following data:

$$\Sigma X_1 = 150 \quad \Sigma X_2 = 98 \quad \Sigma X_3 = 85$$

$$\Sigma X_1^2 = 4634 \quad \Sigma X_2^2 = 1966 \quad \Sigma X_3^2 = 1575$$

$$n_1 = 5 \quad n_2 = 5 \quad n_3 = 5$$

a. What is the value of F_{obt}
b. What is the value of $\overline{X}_G$?
c. What is the value of SS_T?

11. Eight cities were asked to report data on fatalities in auto crashes. The Department of Safety wanted to know if safety devices have an effect on mortality. The department received the following data on the number of fatalities per every 100 crashes where the vehicle was traveling over 40 m.p.h. The type of safety device was also reported. If the data were as follows, answer the questions below.

	Type of Device Seat		
None	Belts	Airbags	Both
16	15	14	6
12	11	9	6
8	7	5	4
10	10	10	5
6	5	4	3
2	1	0	3
9	10	8	6
9	8	6	5

a. What is the value of SS_B?
b. What is the value of $\overline{X}_G$?
c. What is the value of F_{obt}?
d. Would you reject H_0, using $\alpha = .01$?

12. Consider the following data from an independent groups design:

Group 1	Group 2	Group 3
6	5	20
7	4	92
9	4	68
8	3	31
2	1	4
3		10
		82
		212

a. Is it appropriate to use parametric ANOVA to analyze these results? Why or why not?

13. An experiment was conducted to assess the effects of a minor tranquilizer on a performance task at different levels of stress. The levels of stress (Factor A) were moderate and high and the levels of tranquilizer (Factor B) were none and moderate. A two-way ANOVA was done on the data and there was a significant A x B interaction. Explain what this would mean.

Answers: 1. Because the subjects within each group receive the same level of the independent variable, variability among the scores within each group cannot be due to differences in the effect of the independent variable. **2a**. 65.41 **2b**. 588.66 **3**. 82 **4a**. The treatments have an effect on the number of relapses experienced by clients with alcoholism. **4b**. The treatments have no effect on the number of relapses experienced by clients with alcoholism. **4c**. 76 **4d**. 270 **4e**. 346 **4f**. $F_{obt} = 23.68$ **4g**. $F_{crit} = 3.10$; therefore reject H_0 **5**. No, the tests are equivalent and equally powerful. With a little algebra one can derive the t equation from the F equation with k = 2 **6a**. 92.067 **6b**. 6.93 **6c**. 1.96 **6d**. $F_{crit} = 6.93$, $F_{obt} = 1.96$; retain H_0 **7a**. H_1 = the flavor of toothpaste has an effect on the number of cavities. **7b**. H_0 = the flavor of toothpaste has no effect on the number of cavities. **7c**. 8.190 **7d**. 7.190 **7e**. 3.55 **7f**. 1.14 **7g**. $F_{obt} < F_{crit}$; therefore retain H_0. **8a**. 60.50 **8b**. 2 **8c**. 58.585 **8d**. 6.722 **8e**. 8.715 **9a**. 12 **9b**. 1 **9c**. 14.23 **9d**. 3.77 **9e**. $F_{crit} = 4.75$, reject H_0 **10a**. $F_{obt} = 9.18$ **10b**. 22.2 **10c**. 782.4 **11a**. 85.094 **11b**. 7.281 **11c**. 2.09 **11d**. $F_{crit} = 4.57$, therefore retain H_0 **12**. No, the homogeneity of variance assumption is violated. The F test is not robust with unequal n's. **13**. The effect of the drug on performance was different depending on what the level of stress was.

TRUE-FALSE QUESTIONS

T F 1. $s_B^2 + s_W^2 = s_T^2$.

T F 2. $SS_B + SS_W = SS_T$.

T F 3. A value of -1.0 for F_{obt} is never possible if the calculations are done correctly.

T F 4. The alternative hypothesis in an ANOVA can be either directional or nondirectional.

T F 5. The ANOVA is robust regardless of whether or not sample sizes are equal.

T F 6. A t test for independent groups can be used to analyze the results from an experiment with k = 2 groups.

T F 7. $F^2 = t$.

T F 8. A value of F < 1.0 will result in rejection of H_0.

T F 9. The F test for k = 2 is more powerful than the t test.

T F 10. It is always possible to partition the total variability in a simple ANOVA into two parts; the variability within groups and the variability between groups.

T F 11. An ANOVA can only be used when $n_1 = n_2 = \ldots = n_k$.

T F 12. The ANOVA is used to test the hypothesis that the group variances are equal.

T F 13. s_W^2 is sensitive to the effects of the independent variable.

T F 14. The sum of df_W and df_B equals N - 1.

T F 15. If F_{obt} is significant, it is possible to tell which groups differ from which without further analysis.

T F 16. If F < 1.0 we don't even have to bother to look up F_{crit} since the result cannot possibly be significant.

T F 17. If $n_1 \neq n_2 \neq n_k$ then $SS_T \neq SS_W + SS_B$.

T F 18. The analysis of variance is used to test for differences between sample means when $k \geq 2$.

T F 19. $F_{obt} = s_W^2/s_B^2$

T F 20. $F_{obt} = (SS_B / df_B)/(SS_W/df_W)$

T F 21. $F_{obt} = \dfrac{\sigma^2 + \text{effects of the independent variable}}{\sigma^2 + \text{effects of the dependent variable}}$

T F 22. The F distribution is a normal curve.

Answers: 1. F   **2.** T   **3.** T   **4.** F   **5.** F   **6.** T   **7.** F   **8.** F   **9.** F   **10.** T
11. F   **12.** F   **13.** F   **14.** T   **15.** F   **16.** T   **17.** F   **18.** T   **19.** F   **20.** T
21. F   **22.** F.

SELF-QUIZ

1. The total degrees of freedom for an experiment with $n_1 = 10$, $n_2 = 12$, and $n_3 = 10$ is

    _____.

    a. 2
    b. 32
    c. 31
    d. 29

2. Which of the following is(are) illegal values for F_{obt}?

    a.   1.00
    b.   0.96
    c.   -2.97
    d.   b and c

3. What is the value of F_{crit} for an experiment with $k = 3$, $n_1 = n_2 = n_3 = 10$ subjects, and $\alpha = .01$?

    a. 5.45
    b. 4.51
    c. 3.35
    d. 5.49

4. What is the value of F_{obt} for the following data?

Group 1	Group 2	Group 3
4	11	1
9	11	6
10	10	4

    a. 5.67
    b. 6.53
    c. 1.95
    d. 3.26

5. What is the value of SS_T if $SS_B = 236$ and $SS_W = 54$?

    a. 290
    b. 182
    c. 4.37
    d. 100

6. By doing multiple t tests when there are more than 2 experimental groups we increase the risk of making what kind of mistake?

    a. accepting H_0
    b. Type I error
    c. Type II error
    d. all of the above

7. If $s_1^2 = 10$, $s_2^2 = 15$, and $s_3^2 = 12$, and $n_1 = n_2 = n_3$, then s_w^2 equals _____.

   a. cannot be determined from information given
   b. 37.0
   c. 12.33
   d. 3.51

8. If $s_1^2 = 9$, $s_2^2 = 6$, $s_3^2 = 8$, and $n_1 \neq n_2 \neq n_3$, then s_w^2 equals _____.

   a. cannot be determined from information given
   b. 7.667
   c. 2.77
   d. 23.0

9. If $\overline{X}_1 = 46$, $\overline{X}_2 = 50$, $\overline{X}_3 = 92$ and $n_1 = n_2 = n_3$, what is the value of $\overline{X}_G$?

   a. cannot be determined from information given
   b. 50.0
   c. 188.0
   d. 62.67

10. If $SS_W = 126$, $N = 28$, $k = 4$, then what is the value of s_w^2

   a. cannot be determined from information given
   b. 5.25
   c. 42.0
   d. 4.667

11. If $df_B = 3$ and $df_T = 29$ and $F_{obt} = 3.15$, what would you conclude using $\alpha = .05$?

   a. reject H_0
   b. reject H_1
   c. retain H_0
   d. retain H_1

12. If $k = 2$ and $t_{obt} = 2.95$, what would the value of F_{obt} be for an independent groups design?

   a. 2.95
   b. 8.70
   c. 1.72
   d. 0.05

13. If $s_B^2 = 27.9$ and $s_W^2 = 54.2$, what is the value of F_{obt}?

    a. impossible result, there must be an error
    b. 1.94
    c. 0.51
    d. 2.76

14. If $SS_T = 96$, $SS_W = 47$, and $SS_B = 68$, what would you conclude?

    a. reject H_0
    b. retain H_0
    c. cannot be determined from information given
    d. there must be an error in the calculations

15. How many variance estimates are there in a 2 x 2 factorial design?

    a. 1
    b. 2
    c. 3
    d. 4

16. Consider the following graphic results from a 2 x 2 factorial experiment.

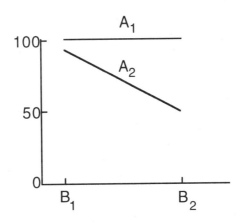

Would you think there is a significant A x B interaction?

    a. yes
    b. no, it is not possible to have a significant interaction in a 2 x 2 factorial experiment.
    c. no, the lines should be parallel for a significant interaction

Answers: **1.** c   **2.** c   **3.** d   **4.** b   **5.** a   **6.** b   **7.** c   **8.** a   **9.** d   **10.** b
11. a   **12.** b   **13.** c   **14.** d   **15.** d   **16.** a.

17 MULTIPLE COMPARISIONS

CHAPTER OUTLINE

I. Introduction

A. <u>Uses of multiple comparison techniques</u>. When ANOVA is used on k > 2 groups, multiple comparisons are made to determine which conditions differ. The F ratio tells whether some conditions do differ but not which ones.

B. *A priori* or planned comparisons.

   1. These comparisons are planned in advance of the experiment and often arise from predictions based on theory or prior research.

   2. Do not correct for higher probability of a Type I error.

   3. Use t test for independent groups except we use s_W^2 as an estimate of σ^2 for the denominator of the t equation. Equation is:

$$t_{obt} = \frac{\overline{X}_1 - \overline{X}_2}{s_W^2 \left(\frac{1}{n_1} + \frac{1}{n_2} \right)}$$

with df = N - k.

If $n_1 = n_2 = n$, the formula simplifies to:

$$t_{obt} = \frac{\overline{X}_1 - \overline{X}_2}{\sqrt{\dfrac{2 s_W^{\,2}}{n}}}$$

If $|t_{obt}| \geq |t_{crit}|$, reject H_0

4. This same t test can be applied for testing any pair of sample means $(\overline{X}_k)$ that was planned prior to the experiment.

5. The number of planned comparisons should be kept to a minimum and flow logically from the experimental design.

C. *A posteriori* or *post hoc* comparisons.

1. These are comparisons between groups not planned prior to the experiment.

2. They maintain the Type I error rate at α while making all possible comparisons between sample means.

3. Sampling distributions for multiple comparisons are called Q or studentized range distributions and are developed by randomly taking k samples of equal n from the same population and determining the difference between the highest and lowest sample means.

4. Error rates.

   a. Experiment-wise error rate is the probability of making one or more Type I errors for the full set of possible comparisons in an experiment.

   b. Comparison-wise error rate is the probability of making a Type I error for any of the possible comparisons.

II. Tukey HSD Test

A. Maintains the experiment-wise Type I error rate at α.

B. Statistic calculated is:

$$Q_{obt} = \frac{\overline{X}_i - \overline{X}_j}{\sqrt{\dfrac{s_w^2}{n}}}$$

where
$\overline{X}_i =$ the larger of the two means being compared
$\overline{X}_j =$ the smaller of the two means being compared
$s_w^2 =$ the within-groups variance estimate
$n =$ the number of subjects in each group

C. Since $\overline{X}_i > \overline{X}_j$, Q_{obt} is always positive.

D. <u>Use Q distribution to evaluate Q_{obt}</u>. Q_{crit} depends on n, k, and α.
If $Q_{obt} \geq Q_{crit}$, reject H_0.

E. Overall F must be significant to apply the HSD or any *post hoc* test.

III. Newman-Keuls Test

A. *Post hoc* test which holds the comparison-wise error rate at α rather than for the entire set of comparisons.

B. Q_{obt} is calculated the same way as for HSD test.

C. Varies the value of Q_{crit} for each comparison.

1. Value of Q_{crit} for any comparison is given by the sampling distribution of Q for the number of groups having means encompassed by $\overline{X}_i$ and $\overline{X}_j$ after all the means have been rank ordered.

2. The number of groups in the comparison is symbolized by r.

3. Different critical value for each comparison depending on r.

D. To use Newman-Keuls test the overall F must be significant.

E. Degrees of freedom equal N - k.

F. **If $Q_{obt} \geq Q_{crit}$, reject H_0.**

G. With unequal n's (provided they are not too different), use the harmonic mean (n) of the various n's and use it in the denominator of the Q equation for both HSD and Newman-Keuls.

$$\tilde{n} = \frac{k}{\dfrac{1}{n_1} + \dfrac{1}{n_2} + \dfrac{1}{n_3} + \cdots + \dfrac{1}{n_k}}$$

where　$k =$　number of groups
$n_k =$　number of subjects in the kth group

IV. Comparison of Tests

A. Planned comparisons more powerful than *post hoc* tests since they do not correct for increased probability of making a Type I error.

　　1. Planned comparisons should be relatively few

　　2. Should flow logically from the experimental design

B. Newman-Keuls has higher experiment-wise Type I error rate than HSD.

C. Newman-Keuls is more powerful than HSD.

D. Newman-Keuls has a lower probability of making Type II error than HSD.

E. Which test to use depends on researcher's assessment of which error rate (Type I or II) to minimize.

CONCEPT REVIEW

Previously we have learned that the ANOVA is a statistical technique to determine if the independent variable has had a significant effect in a multi-group experiment using (1) _____

overall test. But usually we are interested in determining which of the conditions (2) _____ from each other. This is done by making (3) _____ (4) _____ between pairs of group (5) _____. These comparisons may be either *a priori*, sometimes called (6) _____ comparisons or they may be

(1) one

(2) differ

(3) multiple
(4) comparisons
(5) means

(6) planned

a posteriori or (7) _____ (8) _____ comparisons.

Planned comparisons are made in advance of the

experiment. They are generally (9)_____ in (10)_____, and

(9) few
(10) number
(11) theory

arise due to (11) _____. In the planned comparisons we do not

correct for the (12) _____ probability of a Type (13) _____

(12) higher
(13) I

error. There is controversy among statisticians about planned

comparisons but for our purposes these comparisons do not

need to be (14) _____ as long as there are (15) _____ of

(14) independent or
       orthogonal
(15) few
(16) flow logically

them and they (16) _____ from the experimental design.

In doing planned comparisons we use the t test for

(17) _____ (18) _____ except that we use (19) _____ as

(17) independent
(18) groups
(19) s_W^2
(20) σ^2

an estimate of (20) _____ in the denominator. The equation

becomes:

$$t_{obt} = \frac{\overline{X}_i - \overline{X}_j}{\sqrt{(21)\;\underline{\quad}\left(\frac{1}{n_i} + \frac{1}{n_j}\right)}}$$

(21) s_W^2

with (22) _____ degrees of freedom. If $n_1 = n_2 = n$, then the

(22) $N - k$

equation simplifies somewhat to:

$$t_{obt} = \frac{\overline{X}_i - \overline{X}_j}{\sqrt{\frac{(23)\;\underline{\quad}}{n}}}$$

(23) $2s_W^2$

The significance of t_{obt} is evaluated in the usual manner.

If | (24) _____ | ≥ | (25) _____ |, then (26) _____ H$_0$.

(24) t_{obt}
(25) t_{crit}
(26) reject

If comparisons are not planned a different procedure referred

to as (27)_____ analysis is used. Since the comparisons were

(27) *post hoc*

not planned (28)_____ the experiment, we must correct for the

(28) before

(29) _____ probability of making a Type (30) _____ error

(29) higher
(30) I

when comparisons are made for more than (31) _____ means.

(31) two

The sampling distributions for *post hoc* multiple comparisons are

called the (32) _____ or (33) _____ (34) _____ distribu-

(32) Q
(33) studentized
(34) range
(35) k
(36) same

tions. These distributions were developed by taking (35) _____

samples of equal n from the (36) _____ population and

determining the (37) _____ between the highest and lowest

(37) difference

sample (38) _____. The differences were then divided by

(38) means

(39) _____ producing distributions like the (40) _____

(39) $\sqrt{s_w^2/n}$
(40) t
(41) multiple

distributions only these are appropriate for making (41) _____

comparisons.

The two *a posteriori* tests described in this chapter are only

two of many possible tests for multiple comparisons. The tests are

different in how they correct for the increasing probability of mak-

ing a Type I error. The (42) _____-wise error rate is the prob-

(42) experiment

ability of making one or more Type I errors for the full set of poss-

ible comparisons in an experiment. The (43) _____-wise

(43) comparison

error rate is the probability of making a Type I error for

(44) _____ of the possible comparisons.

The Tukey (45) _____ (46) _____ (47) _____ (HSD)

test maintains the (48) _____ -wise Type I error rate at

(49) _____. The statistic calculated for this test is (50) _____.

The formula is:

$$(51)\ \underline{}_{obt} = \frac{(52)\ \underline{}\ -\ (53)\ \underline{}}{\sqrt{\dfrac{(54)\ \underline{}}{(55)\ \underline{}}}}$$

where

$\overline{X}_i$ = the (56) _____ of the two means being compared.

$\overline{X}_j$ = the (57) _____ of the two means being compared.

s_W^2 = the (58) _____ -groups (59) _____ estimate.

n = the (60) _____ of subjects in (61) _____ group

To determine Q_{crit} we enter Table G using the degrees of free-

dom associated with (62) _____, (63) _____ (the number of

groups in the experiment) and (64) _____. Q_{obt} will always

be (65) _____. If Q_{obt} (66) _____ Q_{crit}, reject H_0.

The (67) _____ -(68) _____ test is the other *post hoc*

test presented. It is different from the HSD in that it maintains

the (69) _____ -wise error rate at (70) _____. This means

the Type I error rate is at α for (71) _____ comparison, not for

the (72) _____ set of comparisons. In order to accomplish this,

(44) any
(45) honestly
(46) significant
(47) difference
(48) experimentl
(49) α
(50) Q
(51) Q
(52) $\overline{X}_i$
(53) $\overline{X}_j$
(54) s_W^2
(55) n
(56) larger
(57) smaller
(58) within
(59) variance
(60) number
(61) each
(62) s_W^2
(63) k
(64) α
(65) positive
(66) $\geq$
(67) Newman
(68) Keuls
(69) comparison
(70) α
(71) each
(72) entire

the Newman-Keuls test varies the value of (73) _____ for each

comparison. The value of Q_{crit} for any comparison is derived

from the (74) _____ (75) _____ of Q for the (76) _____

of groups which are encompassed by (77) _____ and

(78) _____ after all the means have been (79) _____ ordered.

The number of means encompassed by $\overline{X}_i$ and $\overline{X}_j$ is symbol-

ized by (80) _____. The value of Q_{crit} depends on

(81) _____, (82) _____, and (83) _____. Q_{obt} is calculat-

ed the same as for the HSD test substituting the appropriate values

of $\overline{X}_i$ and $\overline{X}_j$. Just as for any *post hoc* comparison, to use the

Newman-Keuls test the (84) _____ (85) _____ must be

(86) _____ before proceeding with the multiple comparisons.

   Both of the above *a posteriori* tests are appropriate when there

are (87) _____ n's in each group. If the n's do not differ

greatly we can still use these tests by calculating the

(88) _____ (89) _____ of the various n's and use it in the

(90) _____ of the Q equation. The formula for the harmonic

mean (symbolized by (91) _____) is:

$$\tilde{n} = \frac{(92) \;\rule{1cm}{0.4pt}}{\dfrac{1}{n_1} + \dfrac{1}{n_2} + \dfrac{1}{n_3} + \cdots + \dfrac{1}{(93) \;\rule{0.8cm}{0.4pt}}}$$

where

   (94) _____ = number of groups

(73) Q_{crit}

(74) sampling
(75) distribution
(76) number
(77) $\overline{X}_i$
(78) $\overline{X}_j$
(79) rank

(80) r

(81) df
(82) r
(83) α

(84) overall
(85) F
(86) significant

(87) equal

(88) harmonic
(89) mean
(90) denominator

(91) n

(92) k
(93) n_k

(94) k

n_k = number of (95) _____ in the kth group   (95) subjects

Since (96) _____ comparisons do not correct for   (96) planned

an increased probability of making a Type I error, they

are more (97) _____ than either of the *post hoc* tests. In gen-   (97) powerful

eral, for the *a posteriori* tests, the (98) _____ test is more power-   (98) Newman-Keuls

ful than the (99) _____ test. Newman-Keuls has a higher   (99) HSD

(100) _____-wise error rate, but a lower (101) _____ error   (100) experiment

    (101) Type II

rate. One could say the HSD was the more (102) _____ of the   (102) conservative

two *a posteriori* tests. In deciding which test to use the researcher

would choose the test which (103) _____ the appropriate error   (103) minimized

rate.

EXERCISES

1. A pharmacologist is interested in determining whether or not three different psychoactive drugs differ in how long they remain in the body before they are excreted or metabolized. The pharmacologist randomly assigned 18 subjects to three different groups and recorded the number of days the drugs remained at measurable levels in each subject. The results are presented below.

Drug A	Drug B	Drug C
8	2	3
7	1	3
6	3	1
6	2	4
5	4	3
2	5	4

    a. Is there a difference in the time it takes for these drugs to be cleared from the subjects? (Use $\alpha = .05$).

     b. Which drugs are different in how long they remain in the subjects? (Use $\alpha = .05$ and the HSD test).

     c. Same as 1b, but using the Newman-Keuls test.

2.   A company which sells honey wishes to find out if the strain of bee they use (strain 1) is the most productive bee available. To study the question the four avialable strains are tested at their research farm to see how much honey is produced by each type of bee. Test hives are set up and the number of ounces of honey each hive produces is recorded and given below.

Strain 1	Strain 2	Strain 3	Strain 4
58	56	55	29
62	59	67	42
71	68	66	50
73	68	58	51
80	70	64	48

     a. Are any *a priori* comparisons appropriate and if so which ones?

     b. Using $\alpha = .05_{2 \text{ tail}}$ is there a difference between strain 1 and strain 2 using planned comparisons?

     c. Using $\alpha = .05$ and the HSD test, which strains are different in the amount of honey they produce?

     d. Same as 2c, using the Newman-Keuls test.

3.   A clinical psychologist wants to know if there is any difference in the age of onset of paranoid schizophrenia, simple schizophrenia, or catatonic schizophrenia. The psychologist does a chart review and records the age of diagnosis for random samples of patients with the above diagnoses.

Simple	Paranoid	Catatonic
20	19	20
32	25	22
19	20	18
24	31	30
22	25	25

     a. Is there any difference in the age of onset for these three types of schizophrenia? (use $\alpha = .01$).

     b. Make all *post hoc* pairwise comparisons possible, using $\alpha = .01$ and the Newman-Keuls test.

4. A cancer researcher is interested in the effect of certain drugs on the growth rate of cancer cells. She grows 24 cultures of cancer cells and randomly divides them into four groups of 6 cultures each. Then she injects the cultures of each group with a different drug and counts the number of cells per square millimeter after 7 days. The following data are the results.

Control	Drug X	Drug Y	Drug Z
100	86	99	104
106	79	100	106
108	85	102	110
110	82	101	109
104	90	105	115
101	75	96	120

a. Use one-way ANOVA to determine if any of the drugs has an effect ($\alpha = .05$).

b. Use the Newman-Keuls test to compare the 4 drugs ($\alpha = .05$).

5. A physiological psychologist suspects that the sweat gland activity decreases in people as they get older. The psychologist believes that the highest readings should occur in the youngest people and the lowest readings should occur in the oldest people. The sweat gland activity is assessed by measuring the electrodermal activity of the skin. The following data are electrodermal readings measured in micromhos/cm

	Age Group	
Young	Middle	Old
8.6	8.5	2.1
11.2	9.0	1.0
10.1	10.2	3.6
9.0	7.6	5.4
9.8	9.5	2.7
13.6	7.0	2.6

a. Using an *a priori* test, evaluate whether or not the young group is different from the old group using $\alpha = .05_{1 \text{ tail}}$.

b. Using the Newman-Keuls test, evaluate all possible pair-wise comparisons using $\alpha = .05$. (Assume F_{obt} is significant.)

c. Would the results of part b have been different had we used the HSD test?

6. An experimental psychologist used 5 different methods to teach rats to run a maze. The results of an F test are shown below along with the sample means.

Source	SS	df	s^2	F
Between groups	208.8	4	52.2	7.25*
Within groups	108.0	15	7.2	
Total	316.8	19		

*With α -= .01 F_{crit} = 4.98, therefore reject H_0.

$\overline{X}_1$ = 13.75, $\overline{X}_2$ = 7.75 , $\overline{X}_3$ = 13.25 , $\overline{X}_4$ = 12.50 , $\overline{X}_5$ = 5.75 (seconds)

a. Do a planned comparison test on the difference between group 2 and group 4. Are the results significant using α = .05$_{2\ tail}$? Assume equal n's in each group.
b. What is the value of Q_{crit} at α = .01 for the HSD test?
c. Is group 4 different from group 5 using α = .01 with the Newman-Keuls test?

7. From an experiment the following data are obtained:

$$\overline{X}_1 = 63.00 \quad n_1 = 4$$
$$\overline{X}_2 = 73.25 \quad n_2 = 4$$
$$\overline{X}_3 = 44.60 \quad n_3 = 5$$
$$s_w^2 = 9.595 \quad df\ for\ s_w^2 = 10$$

a. Using α = .01 and the Newman-Keuls test, compare the likelihood that X_1 and X_3 are random samples from the same population.
b. What is the value of n?

Answers: 1a. F_{obt} = 5.24; F_{obt} = 3.68; therefore, reject H_0. There is a difference in the time required for these drugs to be cleared from the body.

1b. Q_{obt} for Drug A and Drug B comparison = 4.09; Q_{crit} = 3.67; therefore reject H_0. Q_{obt} for Drug A and Drug C comparison =3.83; Q_{crit} = 3.67; therefore reject H_0. Q_{obt} for Drug B and Drug C comparison = 0.26; Q_{crit} = 3.67; therefore retain H_0.

1c. All values for Q_{obt} will be the same as for the HSD test only Q_{crit} will change. For comparison of Drug A and B, Q_{crit} = 3.67; therefore reject H_0. For comparison of Drug A and C, Q_{crit} = 3.01; therefore reject H_0. For comparison of Drug B and C, Q_{crit} = 3.01; therefore retain H_0. By both analyses Drug A differs from both drugs B and C in the time required to be cleared from the body. However, we cannot detect a difference between drugs B and C.

2a. Yes, *a priori* comparisons of strain 1 and 2, 1 and 3, 1 and 4 are appropriate.

2b. $t_{obt} = 0.97$; $t_{crit} = \pm 2.120$ therefore retain H_0.

2c. Q_{crit} for all comparisons = 4.05. Q_{obt} for 1 vs 2 = 1.37; retain H_0. Q_{obt} for 1 vs 3 = 2.02; retain H_0. Q_{obt} for 1 vs 4 = 7.36; reject H_0. Q_{obt} for 2 vs 3 = 0.65; retain H_0. Q_{obt} for 2 vs 4 = 6.00; reject H_0. Q_{obt} for 3 vs 4 = 5.34; reject H_0.

2d. Q_{obt} will be same as for 2d only Q_{crit} will change. Q_{crit} for 1 vs 2 = 3.00; retain H_0. Q_{crit} for 1 vs 3 = 3.65; retain H_0. Q_{crit} for 1 vs 4 = 4.05; reject H_0. Q_{crit} for 2 vs 3 = 3.00; retain H_0. Q_{crit} for 2 vs 4 = 3.65; reject H_0. Q_{crit} for 3 vs 4 = 3.00; reject H_0

3a. $F_{obt} = .053$; retain H_0. We cannot detect a difference here.

3b. No *post hoc* comparisons are significant.

4a. $F_{obt} = 39.21$; F_{crit} for 3 and 20 df = 3.10; therefore reject H_0.

4b. Q_{obt} for control vs X = 11.48; $Q_{crit} = 3.58$; reject H_0. Q_{obt} for control vs Y = 2.26; $Q_{crit} = 2.95$; retain H_0. Q_{obt} for control vs Z = 3.04; $Q_{crit} = 2.95$; reject H_0. Q_{obt} for X vs Y = 9.22; $Q_{crit} = 2.95$; reject H_0. Q_{obt} for X vs Z = 14.53; $Q_{crit} = 3.96$; reject H_0. Q_{obt} for Y vs Z = 5.31; $Q_{crit} = 3.58$; retain H_0. Drug X differs significantly from all the others.

5a. $t_{obt} = 8.52$; $t_{crit} = 1.753$; therefore reject H_0.

5b. Q_{obt} young vs middle = 2.82; $Q_{crit} = 3.01$; retain H_0. Q_{obt} young vs old = 12.04; $Q_{crit} = 3.67$; reject H_0. Q_{obt} middle vs old = 9.23; $Q_{crit} = 3.01$; reject H_0.

5c. No, since Q_{crit} would be 3.67 for all tests if we had used the HSD test.

6a. $t_{obt} = 2.50$; $t_{crit} = 2.131$; therefore reject H_0.

6b. $Q_{crit} = 5.56$.

6c. $Q_{obt} = 5.031$; $Q_{crit} = 4.84$; therefore reject H_0.

7a. $Q_{obt} = 12.30$; $Q_{crit} = 4.48$; therefore reject H_0.

7b. 4.29

TRUE-FALSE QUESTIONS

T F 1. It is possible to have an overall F which is significant and have all the *a priori* comparisons be nonsignificant.

T F   2. It is not necessary to do an F test if one is only interested in doing planned comparisons between the groups.

T F   3. The most powerful and most legitimate way of doing multiple comparisons is to plan to do all possible pairwise comparisons prior to the experiment.

T F   4. If $k = 2$, the t test for independent groups, the F test, and the t test for planned comparisons are all essentially the same test.

T F   5. The reason we use s_w^2 based on all the groups instead of just two groups in doing multiple comparisons is because it gives us the best estimate of μ.

T F   6. If $k = 3$, Q_{crit} for the HSD is the same as Q_{crit} for Newman-Keuls if one is testing the difference between the highest and lowest means.

T F   7. The HSD test is less powerful than the planned comparisons.

T F   8. The Newman Keuls test has a lower probability of making a Type II error than the HSD test.

T F   9. The HSD holds the comparison-wise probability of making a Type I error at α.

T F 10. In the HSD test where $k = 3$ if the lowest and highest pair comparison is nonsignificant, none of the other comparisons will be significant either.

T F 11. All researchers agree that in order to do *a priori* comparisons the comparisons do not need to be orthogonal (independent).

T F 12. For the Newman-Keuls test as r increases so does the value of Q_{crit} if α and k are held constant.

T F 13. For both correlation and multiple comparisons r refers to the correlation coefficient.

T F 14. Q_{obt} can never be negative.

T F 15. If for $k = 4$, $n_1 = 10$, $n_2 = 12$, $n_3 = 30$, and $n_4 = 100$; one can still properly apply either the HSD or Newman-Keuls test provided one calculates the harmonic mean (n) to use in the denominator of the Q equation.

T F 16. In general, if the consequence of making a Type I error were much greater than making a Type II, it would be preferable to use the HSD test for a *post hoc* analysis.

T F 17. The probability of making a Type II error is the same, regardless of sample size, for the HSD test.

T F 18. If one did a series of *a priori* pairwise t tests on k > 2 groups and analyzed the results using the t distribution, there would be a substantially greater risk of making a Type II error than if one used HSD or Newman-Keuls tests and the Q distribution.

Answers: 1. F   **2.** T   **3.** F   **4.** T   **5.** F   **6.** T   **7.** T   **8.** T   **9.** F   **10.** T
11. F   **12.** T   **13.** F   **14.** T   **15.** F   **16.** T   **17.** F   **18** F.

SELF-QUIZ

1. The value for Q_{crit} using the HSD test for 6 groups with 6 subjects in each group and α = .01 is _____.

   a. 4.30
   b. 4.23
   c. 5.24
   d. 5.37

2. The harmonic mean when k = 3 and $n_1 = 5$, $n_2 = 4$, and $n_3 = 7$ is _____.

   a. 5.06
   b. 1.69
   c. 0.19
   d. 0.06

3. In general the number of planned comparisons should _____.

   a. equal k
   b. be kept to a minimum
   c. equal n_k
   d. equal zero

4. Which of the following tests hold the experiment-wise error rate at α?

   a. HSD test
   b. Newman-Keuls
   c. F test
   d. a and c

5. In general, which of the following tests is the most powerful test to detect a difference between group means?

    a. a t test for a planned comparison
    b. the HSD test
    c. the Newman-Keuls test
    d. all equally powerful

6. If $n = 8$ and $s_w^2 = 18.2$, what is the value of the denominator for a planned comparison?

    a. 4.55
    b. 2.27
    c. 1.51
    d. 2.13

7. If $\overline{X}_1 = 10.2$, $\overline{X}_2 = 16.1$, $\overline{X}_3 = 12.6$, and $\overline{X}_4 = 9.0$ the value of Q_{crit} for the comparison of $\overline{X}_1$ and $\overline{X}_4$ with $\alpha = .01$ is _____ if s_w^2 has 12 degrees of freedom.

    a. 5.05
    b. 5.50
    c. 4.32
    d. 5.84

For problems 9 - 12 consider the following information from an experiment to compare differences between the length of time a house paint lasts before it begins to fade. The number of subjects are the same in each group.

Source	SS	df	s^2	F
Between groups	720.096	2	360.048	21.54
Within groups	300.857	18	16.714	
Total	1020.950	20		

$$\overline{X}_1 = 21.5714, \overline{X}_2 = 26.5714, \overline{X}_3 = 35.7143, \alpha = .01$$

8. Using the HSD test the value of Q_{obt} for the testing the difference between $\overline{X}_1$ and $\overline{X}_2$ is _____.

   a. 5.92
   b. 2.29
   c. 3.24
   d. 1.05

9. The value of Q_{crit} for the *post hoc* comparison of $\overline{X}_1$ and $\overline{X}_2$ using the HSD test is _____ at $\alpha = .05$.

   a. 2.97
   b. 3.61
   c. 4.00
   d. 4.70

10. The value of Q_{crit} for the *post hoc* comparison of $\overline{X}_1$ and $\overline{X}_2$ using the Newman-Keuls test is _____ at $\alpha = .05$

    a. 2.97
    b. 4.07
    c. 3.61
    d. 4.00

11. The value of t_{obt} for the planned comparison of $\overline{X}_2$ and $\overline{X}_3$ is _____.

    a. 4.96
    b. 5.82
    c. 2.97
    d. 4.18

Answers: 1. c **2.** a **3.** b **4.** d **5.** a **6.** d **7.** c **8.** c **9.** b **10.** a
11. d.

18 INTRODUCTION TO THE TWO-WAY ANALYSIS OF VARIANCE

CHAPTER OUTLINE

I. **Overview**

A. This chapter discusses two-way analysis of variance, independent groups, fixed effects design. With this design the effect of two variables (factors) and their interaction is assessed. The average effect of variable A over the various levels of variable B is called the main effect of variable A. The average effect of variable B over the various levels of variable A is called the main effect of variable B. The interaction of variables A and B is called the interaction effect.

B. Steps in performing the two-way ANOVA.

1. The total sum of squares (SS_T) is partitioned into four components: the within-cells sum of squares (SS_W), the row sum of squares (SS_R), the column sum of squares (SS_C), and the row $\times$ column sum of squares (SS_{RC}).

2. Four variance estimates of σ^2 are formed by dividing each of the above four sum of squares by their degrees of freedom. These estimates are the within-cells variance estimate (s_W^2), the row variance estimate (s_R^2), the column variance estimate (s_C^2), and the row $\times$ column variance estimate (s_{RC}^2).

3. Three F ratios are formed from the four variance estimates. The F ratios are:

$$F_{obt} = s_R^2/s_W^2 \quad \textbf{Main effect of variable A}$$

$$F_{obt} = s_C^2/s_W^2 \quad \textbf{Main effect of variable B}$$

$$F_{obt} = s_{RC}^2/s_W^2 \quad \textbf{Interaction effect of variables A and B}$$

4. The F ratios are evaluated using the following decision rule.

$$\textbf{If } F_{obt} \geq F_{crit}, \textbf{ reject } H_0$$

II Within-cells Variance Estimate (s_W^2)

A. This is equivalent to s_W^2 in one-way ANOVA. Provides an estimate of σ^2. s_W^2 is a measure or the inherent variability of the scores from subject-to-subject. It is based on the variability of the scores within each cell. Therefore, it does not reflect any treatment effect.

B. Equations:

1. Within-cells variance estimate (s_W^2).

$$s_W^2 = SS_W/df_W$$

where SS_W = within-cells sum of squares
df_W = within-cells degrees of freedom

2. Within-cells sum of squares (SS_W).

$$SS_W = SS_{11} + SS_{12} + \cdots + SS_{rc} \quad \textbf{conceptual equation}$$

where SS_{11} = sum of squares for the cell of row 1 and column 1
SS_{rc} = sum of squares for the cell of row r and column c

$$SS_W = \sum x^2 - \left[\frac{\left(\overset{cell\,11}{\sum} x\right)^2 + \left(\overset{cell\,12}{\sum} x\right)^2 + \cdots + \left(\overset{cell\,rc}{\sum} x\right)^2}{n_{cell}} \right] \quad \textbf{computational equation}$$

where $\left(\overset{cell\,rc}{\sum} x\right)^2$ = sum of the scores in the cell of row r and column c, squared

3. Within-cells degrees of freedom (df_W).

$$df_W = rc(n - 1)$$

where r = number of rows
c = number of columns

III. Row Variance Estimate (s_R^2)

A. <u>Measures the main effect of variable A.</u> Based on the differences between row means. Analogous to s_B^2 in one-way ANOVA. It is an estimate of σ^2 + the effects of variable A, averaged over the levels of variable B. If variable A has no effect, then the population row means are equal. Hence:

$$\mu_{a_1} = \mu_{a_2} = \cdots = \mu_{a_r}$$

and the differences among sample row means is just due to random sampling from identical populations.

B. Equations:

1. Row variance estimate (s_R^2).

$$s_R^2 = SS_R/df_R$$

where SS_R = row sum of squares
df_R = row degrees of freedom

2. Row sum of squares (SS_R).

$$SS_R = n_{row}[(\overline{X}_{row\ 1} - \overline{X}_G)^2 + (\overline{X}_{row\ 2} - \overline{X}_G)^2 + \cdots + (\overline{X}_{row\ r} - \overline{X}_G)^2] \quad \textbf{conceptual equation}$$

where $\quad \bar{X}_{\text{row } 1} = \dfrac{\overset{\text{row}}{\underset{1}{\sum}} X}{n_{\text{row } 1}}$

$$\bar{X}_{\text{row } r} = \dfrac{\overset{\text{row}}{\underset{r}{\sum}} X}{n_{\text{row } r}}$$

$$\bar{X}_G = \text{grand mean} = \dfrac{\overset{\text{all scores}}{\sum} X}{N}$$

$$S S_R = \left[\dfrac{\left(\overset{\text{row}}{\underset{1}{\sum}} x\right)^2 + \left(\overset{\text{row}}{\underset{2}{\sum}} x\right)^2 + \cdots + \left(\overset{\text{row}}{\underset{r}{\sum}} x\right)^2}{n_{\text{row}}} \right] - \dfrac{\left(\overset{\text{all scores}}{\sum} x\right)^2}{N} \qquad \begin{array}{l}\textbf{computational} \\ \textbf{equation}\end{array}$$

3. Row degrees of freedom (df_R).

$$\textbf{df}_\textbf{R} = \textbf{r - 1}$$

IV. Column Variance Estimate ($s_C{}^2$).

A. <u>Measures the main effect of variable B.</u> Based on the differences between column means. Analogous to $s_B{}^2$ in one-way ANOVA. It is an estimate of σ^2 + the effects of variable B, averaged over the levels of variable A. If variable B has no effect, then the population column means are equal. Hence:

$$\mu_{b_1} = \mu_{b_2} = \cdots = \mu_{b_c}$$

and the differences among the sample column means are due to random sampling form identical populations.

B. Equations:

1. Column variance estimate ($s_C{}^2$).

$$\textbf{s}_\textbf{C}{}^\textbf{2} = \textbf{SS}_\textbf{C}/\textbf{df}_\textbf{C}$$

where $\quad SS_C$ = column sum of squares

$$df_C = \text{column degrees of freedom}$$

$$SS_C = n_{col.}[(\bar{X}_{col.\ 1} - \bar{X}_G)^2 + (\bar{X}_{col.\ 2} - \bar{X}_G)^2 + \cdots + (\bar{X}_{col.\ c} - \bar{X}_G)^2]\ \textbf{conceptual equation}$$

$$\text{where} \quad \bar{X}_{col.\ 1} = \frac{\displaystyle\sum^{col.\ 1} X}{n_{col.\ 1}}$$

$$\bar{X}_{col.\ r} = \frac{\displaystyle\sum^{col.\ r} X}{n_{col.\ r}}$$

$$SS_C = \left[\frac{\left(\displaystyle\sum^{col.\ 1} X\right)^2 + \left(\displaystyle\sum^{col.\ 2} X\right)^2 + \cdots + \left(\displaystyle\sum^{col.\ c} X\right)^2}{n_{col.}}\right] - \frac{\left(\displaystyle\sum^{all\ scores} X\right)^2}{N} \quad \textbf{computational equation}$$

3. Column degrees of freedom (df_C).

$$df_C = c - 1$$

V. Row × Column Variance Estimate (s_{RC}^2).

A. Measures the interaction effect of variables A and B. An interaction exists when the effect of one of the variables is not the same at all levels of the other variable. Based on the differences among the cell means beyond that which is predicted by the individual effects of the two variables. If there is no interaction and any main effects are removed, then the population cell means are equal. Hence:

$$\mu_{a_1 b_1} = \mu_{a_1 b_2} = \cdots = \mu_{a_r b_c}$$

and differences among cell means must be due to random sampling from identical populations.

B. Equations:

1. Row × column variance estimate (s_{RC}^2).

$$s_{RC}^2 = SS_{RC}/df_{RC}$$

where SS_{RC} = row × column sum of squares

df_{RC} = row × column degrees of freedom

$$SS_{RC} = n_{cell}[(\bar{X}_{cell\ 11} - \bar{X}_G)^2 + (\bar{X}_{cell\ 12} - \bar{X}_G)^2 + \cdots + (\bar{X}_{cell\ rc} - \bar{X}_G)^2]$$

$$- SS_R - SS_C \qquad\qquad \textbf{conceptual equation}$$

where
$$\bar{X}_{cell\ 11} = \frac{\sum\limits^{cell\ 11} X}{n_{cell}}$$

$$\bar{X}_{cell\ rc} = \frac{\sum\limits^{cell\ rc} X}{n_{cell}}$$

$$S S_{RC} = \left[\frac{\left(\sum\limits^{cell\ 11} x\right)^2 + \left(\sum\limits^{cell\ 12} x\right)^2 + \cdots + \left(\sum\limits^{cell\ rc} x\right)^2}{n_{cell}}\right] - \frac{\left(\sum\limits^{all\ scores} x\right)^2}{N} \qquad \begin{array}{l}\textbf{computational}\\\textbf{equation}\end{array}$$

$$- S S_R - S S_C$$

3. Row × column degrees of freedom (df_{RC}).

$$df_{RC} = (r - 1)(c - 1)$$

VI. Computing F Ratios.

A. Equations:

1. To test the main effect of variable A (row effect):

$$F_{obt} = s_R^2/s_W^2$$

2. To test the main effect of variable B (column effect):

$$F_{obt} = s_C^2/s_W^2$$

3. To test the interaction effect of variables A and B (row × column effect):

$$F_{obt} = s_{RC}^2/s_W^2$$

B. Decision Rule.

If $F_{obt} \geq F_{crit}$, reject H_0

This decision rule applies for all three F_{obt} values. Note that it is possible to have all three comparisons significant; none of the comparisons significant; and all possible combinations between these two extremes.

VII.   Interpreting the F comparisons

A. As with one-way ANOVA, a significant F value indicates a real effect. A significant F involving s_R^2 indicates the A variable has had a real main effect. A significant F involving s_C^2 indicates the B variable has had a real main effect. A significant F involving s_{RC}^2 indicates there is a real interaction between A and B. The effects of one or both variables are not the same at all levels of the other variable. If F is not significant, we cannot conclude there has been a real effect, and we must retain H_0. Whichever way we conclude, it is possible we have made a Type I or II error.

VIII.   Multiple Comparisons

A. When conducting a two-way ANOVA, the experimenter is usually interested in more than just determining the main and interaction effects. There is an interest int determining which levels of which variables are having real effects. This is conceptually similiar to Chapter 17 where we took up multiple comparisons in conjunction with one-way ANOVA. However, in the two-way ANOVA it is more complicated and beyond the scope of this textbook.

IX. Assumptions Underlying Two-Way ANOVA

A. The populations from which the samples were taken are normally distributed.

B. Homogeneity of variance.

C. Two-way ANOVA is robust with regard to violations of these assumptions, provided the samples are of equal size

CONCEPT REVIEW

The two-way ANOVA allows us to evaluate in (1) _____   (1) one

experiment the effect of (2) _____ independent variables and the   (2) two

(3) _____ between them. Such experiments are called   (3) interaction

(4) _____ experiments. In a factorial experiment the treatments   (4) factorial

used are (5) _____ of the levels of both variables. In such a design (5) combinations

the effect of variable A averaged over the levels of variable B and

the effect of variable (6) _____ averaged over the levels of var-   (6) B

iable A are called (7) _____ effects. An (8) _____ effect   (7) main
(8) interaction
occurs when the effect of one variable is not the (9) _____ at all   (9) same

levels of the other variable.

In analyzing data from a two-way ANOVA, the total sum of

squares, symbolized by (10) _____ is partitioned into   (10) SS_T

(11) _____ parts, the within sum of squares, symbolized by   (11) four

(12) _____, the row sum of squares, symbolized by   (12) SS_W

(13) _____, the column sum of squares, symbolized by   (13) SS_R

(14) _____, and the row × column sum of squares, symbolized   (14) SS_C

by (15) _____. Thus,   (15) SS_{RC}

(16) SS_R
(17) SS_C
$$SS_T = SS_W + (16) \text{\_\_\_\_} + (17) \text{\_\_\_\_} + (18) \text{\_\_\_\_}$$ (18) SS_{RC}

Each of the sum of squares, SS_W, SS_R, SS_C, and SS_{RC} is used to form a variance estimate of (19) _____, the variance of

the null hypothesis population. The variance estimates are derived by dividing the sum of squares by its (20) _____. Thus,

$$s_W^2 = SS_W/(21) \ _____$$

$$s_R^2 = SS_R/(22) \ _____$$

$$s_C^2 = SS_C/(23) \ _____$$

$$s_{RC}^2 = SS_{RC}/(24) \ _____$$

The within-cells variance estimate, symbolized by (25) _____ is based on the differences among scores (26) _____ each cell and hence is not sensitive to the effects of variables A and B. Thus, it is a measure of σ^2 (27) _____ by the effects of A and B. It is analagous to (28) _____ in the one-way ANOVA. It is the standard against which we assess the effects of (29) _____.

The row variance estimate, symbolized by (30) _____ is based on the differences among the row means. Therefore, it is a measure of σ^2 (31) _____ the effects of variable A. If the F ratio involving s_R^2 is significant, we conclude that variable A has had a real average effect. We say that there is a (32) _____ effect for variable A.

The column variance estimate, symbolized by (33) _____ is based on the differences among the column means. Therefore, it is a measure of σ^2 plus the effects of variable (34) _____. If

(19) σ^2

(20) degrees of freedom

(21) df_W

(22) df_R

(23) df_C

(24) df_{RC}

(25) s_W^2

(26) within

(27) uninfluenced

(28) s_W^2

(29) A and B

(30) s_R^2

(31) plus

(32) main

(33) s_C^2

(34) B

the F ratio involving s_C^2 equals or exceeds (35) _____ we reject   (35) F_{crit}

H_0, and conclude that their was a real (36) _____ effect for   (36) main

variable B.

The row × column variance estimate, symbolized by

(37) _____ is based on the differences among (38) _____   (37) s_{RC}^2
     (38) cell

means beyond that predicted by the individual effects of the two

variables. It is used to evaluate the (39) _____ between var-   (39) interaction

iables A and B. A significant F ratio here indicates that the effect of

one or both of the variables is (40) _____ at all levels of the   (40) not the same

other variable.

Computing the F ratios

The within-cells sum of squares (SS_W) is just the (41) _____   (41) SS

within each cell added together. Thus,

$$SS_W = SS_{11} + SS_{12} + \cdots + (42) _____$$ (42) SS_{rc}

The computational equation for SS_W is:

$$SS_W = \sum_{cell}^{\substack{all \\ scores}} x^2 - \left[\frac{\left(\sum_{11}^{cell} x\right)^2 + \left(\sum_{12}^{cell} x\right)^2 + \cdots + \left(\sum_{rc}^{cell} x\right)^2}{n_{cell}}\right]$$

where $\left(\sum_{rc}^{cell} x\right)^2$ = sum of the scores in the cell of row r and

column c, (43) _____.   (43) squared

The within-cells degrees of freedom is given by:

$$df_W = (44) \underline{\hspace{2cm}}$$

The conceptual equation for the row sum of squares is:

$$SS_R = n_{row}[(\overline{X}_{row\ 1} - \overline{X}_G)^2 + (\overline{X}_{row\ 2} - \overline{X}_G)^2 + \cdots + (\overline{X}_{row\ r} - \overline{X}_G)^2]$$

As the effect of the A variable increases, $(\overline{X}_{row\ r} - \overline{X}_G)^2$ also

(45) \underline{\hspace{2cm}}. Thus SS_R is sensitive to the effect of variable

(45) increases

(46) \underline{\hspace{2cm}}. The computational equation for SS_R is:

(46) A

$$SS_R = \left[\frac{\left(\overset{row\ 1}{\sum} X\right)^2 + \left(\overset{row\ 2}{\sum} X\right)^2 + \cdots + \left(\overset{row\ r}{\sum} X\right)^2}{n_{row}} \right] - \frac{\left(\overset{all\ scores}{\sum} X\right)^2}{N}$$

where $\left(\overset{row\ r}{\sum} X\right)^2$ indicates summing all the scores in row r

and then (47) \underline{\hspace{2cm}}.

(47) squaring

The row degrees of freedom is given by:

$$df_R = (48) \underline{\hspace{2cm}}$$

(48) r - 1

SS_C is very much like SS_R. However, it involves

(49) \underline{\hspace{2cm}} means rather than row means. Hence it is

(49) column

sensitive to the effects of variable B. This can be best seen

from the conceptual equation for SS_C.

$$SS_C = n_{col.}[(\overline{X}_{col.\ 1} - \overline{X}_G)^2 + (\overline{X}_{col.\ 2} - \overline{X}_G)^2 + \cdots + (\overline{X}_{col.\ c} - \overline{X}_G)^2]$$

As the effect of the B variable increases, $(\bar{X}_{col.\ c} - \bar{X}_G)^2$ also

(50) _____. Thus SS_C is sensitive to the effect of variable   (50)   increases

(51) _____. The computational equation for SS_C is: (51)   B

$$SS_C = \left[\frac{\left(\overset{col.1}{\sum} X \right)^2 + \left(\overset{col.2}{\sum} X \right)^2 + \cdots + \left(\overset{col.c}{\sum} X \right)^2}{n_{col.}} \right] - \frac{\left(\overset{all\ scores}{\sum} X \right)^2}{N}$$

where $\left(\overset{col.\ c}{\sum} X \right)^2$ indicates summing all the scores in col. c

and then (52) _____.   (52)   squaring

The column degrees of freedom is given by:

$$df_C = (53) \ _____$$

(53)   c - 1

The row × column sum of squares (SS_{RC}) is a measure of the

differences among (54) _____ means when the individual effects   (54)   cell

of variables A and B have been (55) _____. Thus, it measures   (55)   removed

the (56) _____ between the two variables. The computational   (56)   interaction

equation for SS_{RC} is:

$$SS_{RC} = \left[\frac{\left(\overset{\overset{\text{cell}}{\text{11}}}{\sum} X \right)^2 + \left(\overset{\overset{\text{cell}}{\text{12}}}{\sum} X \right)^2 + \cdots + \left(\overset{\overset{\text{cell}}{\text{rc}}}{\sum} X \right)^2}{n_{\text{cell}}} \right] - \frac{\left(\overset{\overset{\text{all}}{\text{scores}}}{\sum} X \right)^2}{N}$$

$$- SS_R - SS_C$$

where $\left(\overset{\overset{\text{cell}}{\text{rc}}}{\sum} X \right)^2$ indicates summing all the scores in cell rc

and then (57) _____ .

(57) squaring

The row × column degrees of freedom is given by:

$$df_{RC} = (58) \text{_____}$$

(58) (r - 1)(c - 1)

To compute the various F ratios, we first compute the four

(59)_____ and check our computations by computing

(59) sum of squares

(60) _____ . The equation for computing SS_T from the raw

(60) SS_T

scores is:

(61) $\overset{\overset{\text{all}}{\text{scores}}}{\sum} X^2$

$$SS_T = (61) \text{_____} - (62) \text{_____}$$

(62) $\dfrac{\left(\overset{\overset{\text{all}}{\text{scores}}}{\sum} X \right)^2}{N}$

The equation for checking our SS computations is:

(63) SS_R
(64) SS_C

$$SS_T = (63) \text{\_\_\_\_} + (64) \text{\_\_\_\_} + (65) \text{\_\_\_\_} + (66) \text{\_\_\_\_}$$

(65) SS_{RC}
(66) SS_W

Once we are sure our SS computations are correct, we compute the four degrees of freedom. Then we compute the four

(67) _____ . Next we compute the three (68) _____

Finally, we evaluate each F ratio by using the decision rule,.

(67)	variance estimates
(68)	F ratios

If F_{obt} (69) _____ F_{crit}, reject H_0

(69) $\geq$

A significant F value indicates a (70) _____ effect. A sig-

nificant F involving s_R^2 indicates the (71) _____ variable has

had a real main effect. A significant F involving s_C^2 indicates the

(72) _____ variable had a real (73) _____ effect. A significant

F involving s_{RC}^2 indicates there is a real (74) _____ between A

and B. If F is not significant, we must (75) _____ H_0. Whichever

way we conclude, it is possible we have made a (76) _____ or

(77) _____ error.

(70)	real
(71)	A
(72)	B
(73)	main
(74)	interaction
(75)	retain
(76)	Type I
(77)	Type II

To validly use the F test in analyzing the data in the two-way

ANOVA, the populations from which the samples were taken should

be (78) _____ distributed, and there should be (79) _____ of

variance. The F test is (80) _____ with regard to violations of

these mathematical assumptions.

(78)	normally
(79)	homogeneity
(80)	robust

EXERCISES

1. An independent groups experiment is conducted involving variables A and B. There are three levels of each variable, and four subjects randomly assigned to each cell. The data is shown below:

	B		
A	Level 1	Level 2	Level 2
Level 1	2 5 4 7	3 5 6 7	10 8 6 7
Level 2	3 2 4 6	2 6 5 4	8 8 9 6
Level 3	4 7 9 7	5 8 10 7	6 9 7 10

a. What are the null hypotheses for this experiment?

b. Using the two-way ANOVA with $\alpha = .05$, what do you conclude?

2. A sleep researcher is interested in determining whether drinking coffee affects ability to sleep. The researcher has a hunch that it does, but that the effect will depend on how close to bedtime the coffee is drunk. An experiment is conducted in which there are 3 levels of the amount of coffee which is drunk (variable B) and two levels of when the coffee is drunk (variable A). The levels of amount drunk are one, two and three cups of coffee and the coffee is drunk either in the morning or two hours before bedtime. The dependent variable is the number of hours slept that night. Thirty subjects are run in the experiment, with five each being randomly assigned to each cell. The following data are collected.

Time	Coffee		
	1 Cup	2 Cups	3 Cups
Morning	8	9	7
	7	8	9
	8	9	8
	6	6	9
	9	7	8
2 hrs before bedtime	6	5	5
	7	6	4
	8	6	6
	8	6	4
	9	7	5

What do you conclude? Analyze with two-way ANOVA and $\alpha = .05$.

3. A researcher working with elderly people is interested in whether memory decreases with age. An independent groups experiment is conducted with three age groups, 30 - 39, 40 - 49, and 50 - 59. Two memory tasks are given, a difficult, long passage to be memorized and a relatively easy, shorter passage. Subjects are randomly assigned, four to each cell. Scores are percent correct, with 100 being a perfect score. The data are shown below:

Memory Task	Age (Years)		
	30 - 39	40 - 49	50 - 59
Easy	90	88	85
	95	92	94
	97	96	89
	93	91	91
Difficult	80	77	67
	84	75	66
	82	79	69
	79	74	71

a. What are the null hypotheses for this experiment?

b. Analyze these results using two-way ANOVA and $\alpha = .05$. What do you conclude?

4. Complete the following table. There are 5 subjects per cell.

Source	SS	df	s^2	F_{obt}
Rows	1082.358	2		
Columns	2158.364	3		
Row × Column	785.228			
Within-cells	3290.875			
Total				

5. A scientist, interested in weight regulation, believes that if exercise in increased, appetite will increase and more food will be eaten. If this is true, it could reduce the benefits of exercise on weight reduction. A 2×3 factorial experiment is conducted to investigate the effect of exercise on food intake. The experiment involves three levels of exercise and uses male and female rats as subjects. Daily food consumption (grams) is monitored as the dependent variable. The animals are allowed to eat as much food as desired. The following data (grams/day) are obtained:

Gender	Exercise		
	Low	Moderate	High
Female	8	12	14
	10	13	15
	11	11	18
	9	15	15
	10	12	14
Male	13	19	21
	15	12	15
	14	15	16
	14	18	20
	13	17	22

Analyze these results with two-way ANOVA, using $\alpha = .01$. What do you conclude?

Answers: 1a. The A variable has no main effect. The B variable has no main effect. There is no interaction between A and B.

$$\mu_{a_1} = \mu_{a_2} = \cdots = \mu_{a_r}$$

$$\mu_{b_1} = \mu_{b_2} = \cdots = \mu_{b_c}$$

$$\mu_{a_1 b_1} = \mu_{a_1 b_2} = \cdots = \mu_{a_r b_c}$$

1b.

Source	S S	df	s^2	F_{obt}	F_{crit}
Rows (variable A)	30.167	2	15.083	4.60	3.35
Columns (Variable B)	52.667	2	26.334	8.03	3.35
Row × Column	11.667	4	2.917	0.89	2.73
Within-cells	88.50	27	3.278		
Total	183	35			

There is a significant main effect for variable A and variable B. The interaction effect is not significant. Therefore, we reject H_0 for both main effects, but retain H_0 for the interaction between A and B.

2.

Source	S S	df	s^2	F_{obt}	F_{crit}
Rows (Time)	22.533	1	22.533	21.81	4.26
Columns (Coffee amount)	6.200	2	3.100	3.00	3.40
Row × Column	14.467	2	7.233	7.00	3.40
Within-cells	24.800	24	1.033		
Total	68	29			

There is a significant main effect for rows and a significant row × column interaction. There is no significant columns effects. Thus drinking coffee in the morning had no apparent effect., no matter how much was drunk (up to 3 cups, of course). Drinking coffee 2 hours before bedtime interferred with sleep, and the more coffee drunk at this time, the greater the interference.

3a. Memory is the same for the easy and difficult tasks (main effect of variable A). There is no memory change with age (main effect of variable B). There is no interaction between age and memory of material of different levels of difficulty (A × B interaction).

$$\mu_{a_1} = \mu_{a_2} = \cdots = \mu_{a_r}$$

$$\mu_{b_1} = \mu_{b_2} = \cdots = \mu_{b_c}$$

$$\mu_{a_1 b_1} = \mu_{a_1 b_2} = \cdots = \mu_{a_r b_c}$$

3b.

Source	S S	df	s^2	F_{obt}	F_{crit}
Rows (Task difficulty)	1633.500	1	1633.500	200.70	4.41
Columns (Age)	292.000	2	146.000	17.94	3.55
Row × Column	84.000	2	42.000	5.16	3.55
Within-cells	146.500	18	8.139		
Total	2156	23			

There is a significant main effect for task difficulty. There is a significant main effect for age, and there is a significant interaction between age and task difficulty. The "difficult" material was harder to memorize independent of age differences; as age increased there was a loss in memory independent of task difficulty; and the age decrement was more pronounced with the difficult task.

4.

Source	S S	df	s^2	F_{obt}
Rows	1082.358	2	541.179	7.89
Columns	2158.364	3	719.455	10.49
Row × Column	785.228	6	130.871	1.91
Within-cells	3290.875	48	68.560	
Total	7316.825	59		

5.

Source	S S	df	s^2	F_{obt}	F_{crit}
Rows (Female/male)	108.300	1	108.300	26.63	7.82
Columns (Exercise)	140.467	2	70.233	17.27	5.61
Row × column	0.600	2	0.300	0.07	5.61
Within-cells	97.600	24	4.067		
Total	346.967	29			

There is a significant main effect for both rows and columns. The interaction effect is not significant. Thus, the amount of food females and males eat daily is significantly different, independent of exercise level. Increasing the level of exercise increases the amount of food eaten, independent of gender. Exercise level does not appear to have a different effect on males and females regarding the amount of food eaten.

TRUE-FALSE QUESTIONS

T F 1. $s_T^2 = s_R^2 + s_C^2 + s_{RC}^2 + s_W^2$.

T F 2. $SS_T = SS_R + SS_C + SS_{RC} + SS_W$.

T F 3. In general, the alternative hypothesis for two-way ANOVA can be directional.

T F 4 s_W^2 is sensitive to the effects of variable A or B.

T F 5. s_C^2 is a measure of only σ^2, regardless of whether the variables have a main effect.

T F 6. s_R^2 is sensitive to the effects of variable A.

T F   7. The sum of df_R, df_C, df_{RC}, and df_W is N - 1.

T F   8. Two-way ANOVA is like two one-way ANOVA experiments, plus it gives us the ability to analyze the interaction of the two variables.

T F   9. A 2×4 factorial experiment has 2 levels of one variable and 4 levels of the other variable.

T F   10 It is not possible to have an interaction effect unless one of the variables also has a main effect.

T F   11 A main effect refers to the average effect of the variable.

T F   12 In a two-way ANOVA, we do three F tests.

T F   13 s_W^2 is the standard for evaluating the main and interaction effects.

T F   14 In two-way ANOVA, the variability within each cell is used to produce an estimate of σ^2 which is independent of any treatment effects.

T F   15 The variability of cell means is used as a basis of both the interaction effect and the within-cell variance estimate.

T F   16 The assumptions underlying two-way ANOVA and one-way ANOVA are different.

T F   17 In general, when an experimenter does a two-way ANOVA, he has no interest in which levels of the variable have a real effect.

T F   18 In two-way ANOVA, it is possible for all or none, or any combination in between, of the F values to be significant.

T F   19 An assumption of two-way ANOVA is that the sample cell variances are equal.

T F   20 An assumption of two-way ANOVA is that the sample column and row scores are normally distributed.

Answers: **1.** F   **2.** T   **3.** F   **4.** F   **5.** F   **6.** T   **7.** T   **8.** T   **9.** T   **10.** F   **11.** T   **12.** T   **13.** T   **14.** T   **15.** F   **16.** F   **17.** F   **18.** T   **19.** F   **20.** F.

SELF-QUIZ

For questions 1 - 9, use the following data, collected from an independent groups design. $\alpha = .05$.

Variable A	Variable B		
	1	2	3
1	5	4	4
	2	6	9
	1	3	7
	4	1	5
	2	3	4
2	6	6	5
	2	7	7
	3	4	5
	3	3	4
	2	3	8

1. The value of F_{obt} for evaluating the row effect is _____.

   a. 0.66
   b. 6.23
   c. 0.29
   d. 5.28

2. The value of F_{crit} for evaluating the row effect is _____.

   a. 3.40
   b. 7.82
   c. 4.26
   d. 5.61

3. The conclusion regarding the main effect of variable A is _____.

   a. Retain H_0. We cannot conclude variable A has main effect.
   b. Accept H_0. We cannot conclude variable A has main effect.
   c. Reject H_0. Variable A has a significant main effect.
   d. Reject H_0. Variable A has no effect.

4. The value of F_{obt} for evaluating the column effect is _____.

    a. 0.66
    b. 6.23
    c. 0.29
    d. 5.28

5. The value of F_{crit} for evaluating the column effect is _____.

    a. 3.40
    b. 7.82
    c. 4.26
    d. 5.61

6. The conclusion regarding the main effect of variable B is _____.

    a. Retain H_0. We cannot conclude variable B has main effect.
    b. Accept H_0. We cannot conclude variable B has main effect.
    c. Reject H_0. Variable B has a significant main effect.
    d. Reject H_0. Variable B has no effect.

7. The value of F_{obt} for evaluating the row $\times$ column effect is _____.

    a. 0.66
    b. 6.23
    c. 0.29
    d. 5.28

8. The value of F_{crit} for evaluating the row effect is _____.

    a. 3.40
    b. 7.82
    c. 4.26
    d. 5.61

9. The conclusion regarding the interaction effect of variables A and B is _____.

    a. Retain H_0. We cannot conclude there is a significant interaction
    b. Accept H_0. There is no interaction effect between A and B
    c. Reject H_0. There is a significant interaction effect.
    d. Reject H_0. There is no intertaction effect between A and B

10. Consider the following graphic results from a 2 $\times$ 2 factorial experiment. These results show _____.

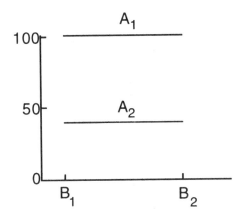

a. there are no significant main effects or interaction effects
b. there is a significant main effect for factor A, no other significant effects
c. there is a significant main effect for factor B, no other significant effects
d. there is a significant interaction effect, no other significant effects

11. Consider the following graphic results from a 2 × 2 factorial experiment. These results show _____.

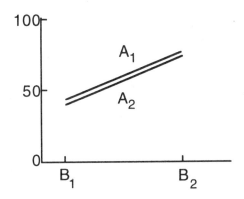

a. there are no significant main effects or interaction effects
b. there is a significant main effect for factor A, no other significant effects
c. there is a significant main effect for factor B, no other significant effects
d. there is a significant interaction effect, no other significant effects

12. Consider the following graphic results from a 2 × 2 factorial experiment. These results show _____.

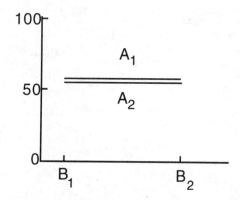

a. there are no significant main effects or interaction effects
b. there is a significant main effect for factor A, no other significant effects
c. there is a significant main effect for factor B, no other significant effects
d. there is a significant interaction effect, no other significant effects

13. Consider the following graphic results from a 2 × 2 factorial experiment. These results show _____.

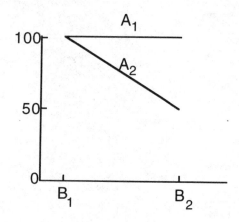

a. there are no significant main effects or interaction effects
b. there is a significant main effect for factor A, no other significant effects
c. there is a significant main effect for factor B, no other significant effects
d. there is a significant interaction effect, no other significant effects

14. Consider the following graphic results from a 2 × 2 factorial experiment. These results show _____.

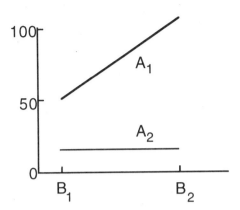

a. there is a significant main effect for factor A, no other significant effects
b. there is a significant main effect for factor B, no other significant effects
c. there is a significant interaction effect, no other significant effects
d. there is a significant main effect for factor A, a significant interaction effect, and no other significant effects
e. there is a significant main effect for factor B, a significant interaction effect, and no other significant effects

Answers: **1.** a **2.** c **3.** a **4.** b **5.** a **6.** c **7.** c **8.** a **9.** a **10.** b **11.** c **12.** a **13.** d **14.** d.

19 | CHI-SQUARE AND OTHER NONPARAMETRIC TESTS

CHAPTER OUTLINE

I. Distinctions Between Parametric and Nonparametric Tests

A. Parametric tests (e.g., t, z, F) depend substantially on population characteristics or parameters for their use.

B. Nonparametric tests (e.g., sign test, Mann-Whitney U test) depend minimally on population characteristics.

C. <u>Distribution free tests.</u> Whereas parametric tests may require that samples be random from normally distributed populations, nonparametrics require that samples be random from populations with the same distributions, hence the term distribution free tests.

D. Advantages for parametric tests.

   1. Parametric tests are generally more powerful and versatile.

   2. Parametric tests are generally robust to violations of the test assumptions.

E. Examples of nonparametric tests.

   1. Sign test

    2. Mann-Whitney U test

    3. Chi-square test

  4. Wilcoxin matched-pairs signed ranks test

II. Chi-Square (χ^2) Single Variable Experiments

A. Often used with nominal data.

B. Allows one to test if the observed results differ significantly from the results expected if H_0 were true.

C. Computational formula

$$\chi^2_{obt} = \sum \frac{(f_o - f_e)^2}{f_e}$$

    where f_o = the observed frequency in the cell
                f_e = the expected frequency in the cell (if H_0 were true)
                $\sum$ = summation over all cells

D. Evaluation of χ^2_{obt}

    1. Family of curves

    2. Vary with df

    3. Lower df curves are positively skewed

    4. k - 1 degrees of freedom where k equals the number of groups or categories

    5. The larger the discrepancy between the observed and expected results the larger the value of χ^2_{obt} and therefore the more unreasonable that H_0 is true.

    6. **If $\chi^2_{obt} \geq \chi^2_{crit}$, reject H_0**

III. Chi-square: Test of Independence Between Two Variables

A. Used to determine whether two variables are related.

B. <u>Contingency table.</u> This is a two-way table showing the contingency between two variables where the variables have been classified into mutually exclusive categories and the cell entries are frequencies.

C. Null hypothesis states that the observed frequencies are due to random sampling from a population in which the proportions in each category of one variable are the same for each category of the other variable.

D. Alternative hypothesis is that these proportions are different.

E. Calculation of χ^2 for contingency tables:

$$\chi^2_{obt} = \sum \frac{(f_o - f_e)^2}{f_e}$$

1. f_e can be found by multiplying the marginals (i.e. row and column totals lying outside the table) and dividing by N.

2. Sum $(f_o - f_e)^2/f_e$ for each cell.

F. Evaluation of χ^2_{obt}.

1. Degrees of freedom for experiments involving the contingency between two variables are equal to the number of f_o scores that are free to vary while at the same time keeping the column and row marginals the same. In equation form:

$$df = (r - 1)(c - 1)$$

where r = number of rows in the contingency table
c = number of columns in the contingency table

2. **If $\chi^2_{obt} \geq \chi^2_{crit}$, reject H_0**

G. Assumptions underlying χ^2

1. Independence exists between each observation in the contingency table.

2. Sample size is large enough so that the expected frequency in each cell is at least 5 for tables where r or c is greater than 2.

3. If table is 1 x 2 or 2 x 2 then each expected frequency should be at least 10.

4. χ^2 can be used with any type of scaling if the data are reduced to mutually exclusive categories and frequency entries.

IV. Wilcoxin Matched-Pairs Signed Ranks Test

A. Used in correlated groups designs with data that are at least of ordinal scaling.

B. Relatively powerful and can be used when assumptions of t test for correlated groups are violated. More powerful than sign test, less powerful than t test.

C. Considers both magnitude and direction of the rank order of the difference scores.

D. Alternative hypothesis stated with no population parameters; e.g. independent variable affects dependent variable.

E. Null hypothesis stated with no population parameters; e.g. independent variable has no effect on dependent variable.

F. Calculation of statistic T_{obt}.

1. Calculate the difference between each pair of scores.

2. Rank the absolute values of the difference scores from the smallest to the largest.

3. Assign to the resulting ranks the sign of the difference score whose absolute value yielded that rank.

4. Compute the sum of the ranks separately for the positive and negative signed ranks. The lower sum is T_{obt}.

5. As a check, the sum of the unsigned ranks should equal $n(n + 1)/2$.

6. If rows scores are tied such that the difference of the paired scores equals zero, then these scores are discarded and N reduced by one.

7. If ties occur in the difference scores, the ranks are given a value equal to the mean of the tied ranks.

G. Evaluation of T_{obt}.

1. **If $T_{obt} < T_{crit}$, reject H_0.** T_{crit} depends on α and N.

H. Assumptions of the signed ranks test.

1. Raw scores must be of at least ordinal scaling.

2. Difference scores must also be of at least ordinal scaling.

V. Kruskal-Wallis Test

A. Used in independent groups design as a substitute for parametric ANOVA when its assumptions are seriously violated. Like parametric ANOVA, Kruskal-Wallis is a nondirectional test.

B. It is a nonparametric test which requires only ordinal scaling of the dependent variable. Does not require population normality.

C. Computes the sum of ranks for each group and tests whether these sums are significantly different. Makes no prediction about population means.

D. Calculation of statistic H_{obt}.

1. Combine all the scores and rank order them, beginning with 1 for the lowest score.

2. Sum the ranks for each group.

3. Substitute these values into the equation and compute H_{obt}.

4. Equation:

$$H_{obt} = \left[\frac{12}{N(N+1)}\right]\left[\frac{R_1^2}{n_1} + \frac{R_2^2}{n_2} + \frac{R_3^2}{n_3} + \cdots + \frac{R_k^2}{n_k}\right] - 3(N+1)$$

where R_1 = sum of the ranks for sample 1
R_2 = sum of the ranks for sample 2
R_3 = sum of the ranks for sample 3
R_k = sum of the ranks for sample k
k = number of samples or groups

E. Evaluation of H_{obt}.

If $H_{obt} \geq H_{crit}$, reject H_0

with H_{crit} found in Table H using df = k-1.

CONCEPT REVIEW

A (1) _____ inference test is one which depends substantial-

ly on population characteristics for its use. Tests which require

only minimal knowledge about the population characteristics
are called (2) _____ tests. These tests are also called

(3) _____ (4) _____ tests since the samples do not have

to be drawn from normally distributed populations. Nonparametric

inference tests have fewer requirements or assumptions about

(5) _____ characteristics. However, they are not used in place of

parametric tests all the time because many parametric tests are

(6) _____ with regard to violations of underlying assumptions.

The main advantages of (7) _____ tests are that they are more

(8) _____ and (9) _____ than nonparametric tests. As a

general rule, an investigator will use (10) _____ tests when-

ever possible.

The test most often employed with (11) _____ data is the

nonparametric test called chi-square. It is symbolized

(12) _____. With this type of data, observations are grouped

into several discrete, (13) _____ (14) _____ categories and

one counts the (15) _____ of occurrence in each category.

The χ^2 test allows one to evaluate if the frequency observed

in each category is significantly (16) _____ from the

frequency (17) _____ in each category if sampling were

(1) parametric

(2) nonparametric

(3) distribution
(4) free

(5) population

(6) robust

(7) parametric

(8) powerful
(9) versatile
(10) parametric

(11) nominal

(12) χ^2

(13) mutually
(14) exclusive
(15) frequency

(16) different

(17) expected

(18) _____ from the (19) _____ (20) _____

Population.

   In order to calculate χ^2 we must first determine the (21) _____

we would expect in each cell. This is symbolized by (22) _____.

The frequency observed is symbolized (23) _____. The closer

the (24) _____ frequency of each (25) _____ is to the

expected frequency for that cell, the more reasonable is

(26) _____. The greater the difference between (27) _____

and (28) _____, the more reasonable (29) _____ becomes.

   To calculate χ^2 the difference between f_o and f_e is

(30) _____ and divided by (31) _____ and then (32) _____

over all of the (33) _____. In equation form this is expressed:

$$\chi^2_{obt} = \sum \frac{((34)\ \_\_\_ - (35)\ \_\_\_)^2}{(36)\ \_\_\_}$$

(18) random
(19) Null
(20) Hypothesis

(21) frequency

(22) f_e

(23) f_o

(24) observed
(25) cell

(26) H_0
(27) f_o
(28) f_e
(29) H_1

(30) squared
(31) f_e
(32) summed
(33) cells

(34) f_o
(35) f_e
(36) f_e

The summation is over (37) _____ the cells. In the case of a

single variable experiment we calculate f_e based on information

about the Null Hypothesis Population. For the experiment with a

single variable there are (38) _____ degrees of freedom. The

decision rule for evaluating the null hypothesis states:

(37) all

(38) k - 1

If χ^2_{obt} (39) _____ χ^2_{crit}, reject H_0

(39) $\geq$

The χ^2 test is a (40) _____ test. Since each cell difference

(40) nondirectional

(41) _____ to the value of χ^2_{obt}, the critical region for rejection   (41) adds

always lies under the (42) _____ hand tail of the χ^2 distribution.   (42) right

One of the main uses of the χ^2 test is determining whether

(43) _____ variables are related. This is done by using a   (43) two

(44) _____ table. This is a (45) _____ -way table showing the   (44) contingency
(45) two
(46) _____ between two variables where the variables have   (46) contingency

been classified into (47) _____ (48) _____   (47) mutually
(48) exclusive
categories and the cell entries are (49) _____. The null hypo-   (49) frequencies

thesis states that there is no (50) _____ between variables   (50) contingency

in the population. If H_1 is true then proportions on one of the var-

iables should be (51) _____ for different categories of the   (51) different

other variable.

The calculation of χ^2_{obt} for a contingency table uses the same

formula as shown above. The most difficult task is to calculate the

value of (52) _____ for each cell. Since we do not know the pop-   (52) f_e

ulation proportions, we (53) _____ them from the (54) _____.   (53) estimate
(54) sample
The value of f_e can be found by (55) _____ the (56) _____ for   (55) multiplying
(56) marginals
that cell and dividing by (57) _____. The marginals are the   (57) N

(58) _____ and (59) _____ totals. If the calculation of f_e is   (58) row
(59) column
correct then the row and column (60) _____ of f_e should equal   (60) totals

the row and column (61) _____. Once f_e is calculated then   (61) marginals

χ^2_{obt} can be calculated using the formula shown above.

The evaluation of χ^2_{obt} is the same as in the one variable case except that there are (62) _____ degrees of freedom for χ^2_{crit}

(62) $(r - 1)(c - 1)$

This is because the degrees of freedom are equal to the number

of (63) _____ scores that are free to (64) _____, while

(63) f_o
(64) vary

keeping the totals (65) _____. Again the decision rule is:

(65) constant

If χ^2_{obt} (66) _____ χ^2_{crit}, reject H_0

(66) $\geq$

The basic assumption in using χ^2 is that there is (67) _____

(67) independence

between each observation recorded in the contingency table. It

is also necessary that the value of (68) _____ is at least 5 for

(68) f_e

the tables where r or c is greater than (69) _____. χ^2 can

(69) 2

be used when data are of (70) _____ type of scaling so long

(70) any

as the data are reduced to (71) _____ (72) _____ categories

(71) mutually
(72) exclusive

and there are appropriate (73) _____.

(73) frequencies

The Wilcoxin signed ranks test is used with the (74) _____

(74) correlated

(75) _____ design with data that are of at least (76) _____

(75) groups
(76) ordinal

scaling. It is more powerful than the (77) _____ test but less

(77) sign

powerful than the (78) _____ test for correlated groups. The

(78) t

statistic calculated is (79) _____. Determining this statistic

(79) T_{obt}

involves four steps:

1. Calculate the (80) _____ between each (81) _____

(80) difference

    of scores.

(81) pair

2. Rank the (82) _____ values of the difference scores

(82) absolute

    from (83) _____ to (84) _____.

(83) smallest
(84) largest

3. Assign the resulting ranks the (85) _____ of the differ-

(85) sign

ence score whose absolute value yielded that rank.

4. Compute the (86) _____ of the ranks separately for the   (86) sum

positive and negative signed ranks. The (87) _____   (87) lower

sum is T_{obt}.

To evaluate T_{obt} the decision rule is: .

If T_{obt} (88) _____ T_{crit}, reject H_0   (88) $\leq$

The Wilcoxin signed ranks test takes into account the (89) _____   (89) rank

(90) _____ of the difference scores not the actual (91) _____   (90) order

of the difference scores.   (91) magnitude

In calculating T_{obt} tied scores are possible. If the raw scores

yield a difference of (92) _____ these scores are disregarded   (92) 0

and the overall (93) _____ is reduced by (94) _____. If ties   (93) N

occur in the difference scores the ranks of these scores are   (94) 1

given a value equal to the (95) _____ of the tied ranks.   (95) mean

The assumptions underlying the Wilcoxin signed ranks test are

that the scores within each pair must be of at least (96) _____   (96) ordinal

scaling and that the (97) _____ scores must also have at least   (97) difference

(98) _____ scaling.   (98) ordinal

The (99) _____ test is used as a substitute for the independ-   (99) Kruskal-Wallis

ent groups, one-way parametric ANOVA. It requires that the data

be of (100) _____ scaling and does not require population   (100) ordinal

(101) _____. It is a (102) _____ test. The statistic used is   (101) normality

(103) _____. Essentially, Kruskal-Wallis tests whether the   (102) nonparametric

          (103) H_{obt}

sums of the (104) _____ for the k samples are so different   (104) ranks

that it is unreasonable to believe that the samples were randomly

selected from populations with identical (105) _____. To   (105) distributions

compute H_{obt}, all the scores are (106) _____ and rank   (106) combined

(107) _____. Next the (108) _____ are summed for each   (107) ordered
                                                                      (108) ranks
sample. It is these (109) _____ of ranks which are evaluated.   (109) sums

The equation for computing H_{obt} is:

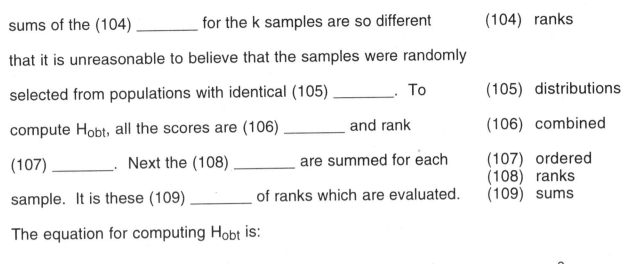

$$H_{obt} = \left[\frac{12}{N(N+1)} \right]\left[\frac{(110)}{(111)} + \frac{R_2^{\,2}}{n_2} + \frac{R_3^{\,2}}{n_3} + \cdots + \frac{R_k^{\,2}}{n_k} \right]$$

$$- 3(N+1)$$

(110) $R_1^{\,2}$
(111) n_1

EXERCISES

In the problems below, if f_e has a decimal remainder, round it to one decimal place in calculating χ^2.

1. A designer of electronic equipment wants to develop a calculator which will have market appeal to high school students. Past marketing surveys have shown that the color of the numeric display is important in terms of market preference. The designer makes up 210 sample calculators and then has a random sample of students from the area high schools rate which calculator they prefer. The calculators are identical except for the color of the display. The results of the survey were that 96 students preferred red, 82 preferred blue, and 32 preferred green.

    a. State H_1 for this experiment.
    b. State H_0 for this experiment.
    c. What is the value of χ^2_{obt}
    d. What do you conclude using $\alpha = .01$?

2. One of the important assumptions underlying the use of parametric statistics is that the sample is randomly selected from a normally distributed population. Consider a sample of N = 500. A sample mean and standard deviation is calculated and we find that the following is true. Between the mean and -1s there are 150 scores. Between the mean and +1s there are 130 scores. Between -1s and -2s there are 70 scores.

Between +1s and +2s there are 82 scores. Beyond +2 standard deviations are 30 scores. Beyond -2 standard deviations are 38 scores.

a. If the population from which this sample was selected were normally distributed, what would the expected frequencies be for each cell in a sample of size 500?

b. Using $\alpha = .01$ what would you conclude about the population from which this sample was selected?

3. A family therapist in a hospital wanted to know if patients with a terminal illness wanted to be informed of their true medical condition. The therapist also wondered if a person age had an effect on their attitude. Because of ethical constraints the therapist asked a healthy sample of subjects who were visitors to the hospital whether they would wish to be told if they had a terminal illness. The age of the respondents was also recorded. The results are shown in the table below.

		Attitude		
		Wanted to be informed	Did not want to be informed	Not sure
Age	< 21	90	15	18
	22 - 35	56	24	19
	36 - 55	50	40	30
	> 55	47	60	50

a. Draw a table with the values of f_e in each of the appropriate cells.
b. What is the value of χ^2_{obt}
c. What is the value of χ^2_{crit} for $\alpha = .01$?
d. What do you conclude?

4. A neuropsychologist wants to determine if people who have a dominant right cerebral hemisphere differ from people with a dominant left cerebral hemisphere in their choice of either music or reading as a preferred activity. He surveyed 127 subjects with the following results.

| | Dominant Hemisphere | |
	Left	Right
Music	48	18
Reading	35	26

a. What is the value of χ^2_{obt}?
b. What is the value of χ^2_{crit} for $\alpha = .05$?
c. What do you conclude?
d. What type error might one be making?

5. A social scientist wants to know if education and socioeconomic status (SES) are independent. He collects the following data.

| | Education | | | |
	No high School	High School	College	Graduate School
High SES	12	19	30	13
Low SES	31	26	32	9

a. What do you conclude using $\alpha = .05$?

6. Consider the following table.

| | | Variable X | |
		Category 1	Category 2
Variable Y	Category 1	6	4
	Category 2	3	5

What is the appropriate statistical test to use to analyze this data if it were all nominal data.

7. A group of pain researchers want to test the hypothesis that different religious groups have different pain complaints. The following data were collected from a review of the patient charts from a hospital pain clinic.

	W	X	Y	Z
Low back pain	9	16	18	40
Headache pain	12	19	36	18
Gastrointestinal pain	20	18	16	21

a. State H_1.
b. State H_0.
c. What do you conclude using $\alpha = .01$?

8a. Given the following data:

		Eyes	
		Blue	Brown
Hair	Blond	36	29
	Brown	23	49

a   Do you think hair color and eye color are independent (use $\alpha = .01$)?
b. What type of error might you be making?

9. A psychologist wants to investigate whether there might be a relationship between birth complications and the development of schizophrenia. In a longitudinal study she gathers the following data.

	Complicated birth	Uncomplicated birth
Schizophrenia	43	15
No Schizophrenia	129	416

a. State H_1.
b. State H_0.
c. What do you conclude using $\alpha = .05$?

10. A new drug is supposed to be effective in reducing motion sickness in people who are prone to such illness. A group of subjects are given a placebo and taken for a ride in a car over a preplanned route. At the end of the trip the subjects are asked to rate their illness on a 20 point scale. A week later the same subjects are given the new drug and taken for an identical ride and asked to rate their degree of illness again. The data are:

Subject	Placebo	Drug
1	16	12
2	20	20
3	18	13
4	10	8
5	15	14
6	12	13
7	15	20
8	19	7
9	17	5
10	13	10

a. State H_1.
b. State H_0.
c. What is the value of T_{obt}?
d. What do you conclude using $\alpha = .05_{2\ tail}$?

11. A group of clients requesting marital therapy were given communication skills training and then rated by independent observers before and after therapy on their ability to resolve problems in a series of hypothetical conflict situations. The results are shown below. A higher score indicates better performance on the task.

Couple	Before	After
1	56	71
2	44	66
3	39	55
4	72	92
5	43	45
6	50	90
7	61	68
8	78	70

    a. What is the value of T_{obt}?

    b. What do you conclude using $\alpha = .05_{2\ tail}$?

12. A political advisor believes that his candidate should not spend time addressing groups of voters who have a low opinion of him. The advisor reasons if they have a low opinion the voter won't change his mind anyway. To test this idea the advisor gets a group from an audience to rate the candidate before and after a speech. From this group he selects a sample of voters who initially rated the candidate poorly and then analyzes the effect of the speech. Here are the data. A higher rating indicates a higher opinion.

Subject	Before	After
1	11	12
2	8	1
3	4	13
4	5	6
5	15	4
6	3	1
7	6	7
8	9	6
9	12	2
10	14	0

    a. What is the value of T_{obt}?

    b. What do you conclude using $\alpha = .05_{1\ tail}$?

    c. What type error might one be making?

13. The following data were collected in an independent groups experiment to test the effect of different levels of a drug on blood pressure (mmHg). Assume the data

seriously violate the assumptions underlying parametric ANOVA. Therefore , you will have to use an alternative test to analyze the data.

Drug		
Level 1	Level 2	Level 3
83	79	90
82	86	80
82	72	70
85	73	65
87	70	78

a. What test will you use?

b. What is your conclusion? Use $\alpha = .01$

14. In an independent groups experiment, four wines are rated by individuals according to taste preference. The resulting data are shown below. The rating scale is from 1 to 20, with 20 representing the highest possible score. Assume the data preclude use of parametric ANOVA because of assumption violations. Analyze these results using an alternative test.

Wine 1	Wine 2	Wine 3	Wine 4
2	4	5	8
3	10	9	13
6	6	12	15
3	7	17	19
1	11	16	14
5	8	20	18

a. What test will you use?

b. Using $\alpha = .05$, what is your conclusion?

Answers: 1a. H_1: there is a difference in color preference for the calculators.   **1b.** H_0: there is no difference in color preference for the calculators.   **1c.** $\chi^2_{obt} = 32.34$ **1d.** $\chi^2_{crit} = 9.210$; therefore reject H_0.

2a. f_e is shown in parentheses.

< -2s	-2s and -1s	-1s and mean	+1s and mean	+1s and +2s	> +2s
38 (11.4)	70 (68.0)	150 (170.6)	130 (170.6)	82 (68.0)	30 (11.4)

2b. $\chi^2_{obt} = 107.50$; $\chi^2_{crit} = 15.086$; therefore reject H_0. It does not seem reasonable to assume that this sample was randomly selected from a normally distributed population.

3a. Table of f_e.

59.9	34.3	28.8
48.2	27.6	23.2
58.4	33.4	28.1
76.5	43.7	36.8

3b. $\chi^2_{obt} = 57.36$ **3c.** $\chi^2_{crit} = 16.812$ **3d.** Attitude and age group are not independent.

4a. $\chi^2_{obt} = 3.30$ **4b.** $\chi^2_{crit} = 3.841$ **4c.** $\chi^2_{obt} < \chi^2_{crit}$; therefore retain H_0. It appears as if preference for music or reading and hemisphere dominance are independent. **4d.** Type II

5a. $\chi^2_{obt} = 7.01$; $\chi^2_{crit} = 7.815$; therefore retain H_0. SES and education appear to be independent.

6. Fisher's exact probability test.

7a. H_1: religion and area of pain are not independent.   **7b.** H_0: religion and area of pain are independent .   **7c.** $\chi^2_{obt} = 25.14$; $\chi^2_{crit} = 16.812$; therefore reject H_0.

8a. $\chi^2_{obt} = 7.65$; $\chi^2_{crit} = 6.635$; therefore it appears as if hair and eye color are not independent.   **8b.** Type I.

9a. H_1: birth complications and occurrence of schizophrenia are not independent.   **9b.** H_0: birth complications and occurrence of schizophrenia are independent.   **9c.** χ^2_{obt} = 65.49; χ^2_{crit} = 3.841; therefore reject H_0.

10a. H_1: the new drug affects self rating of motion sickness.   **10b.** H_0: the new drug has no effect on self rating of motion sickness.   **10c.** T_{obt} = 8.0   **10d.** T_{crit} = 5; therefore retain H_0.

11a. T_{obt} = 3   **11b.** T_{crit} = 3; therefore reject H_0.

12a. T_{obt} = 13   **12b.** T_{crit} = 10; therefore retain H_0.   **12c.** Type II.

13a. Kruskal-Wallis   **13b.** H_{obt} = 3.84; H_{crit} = 9.210. Therefore, retain H_0.

14a. Kruskal-Wallis   **14b.** H_{obt} = 15.10; H_{crit} = 5.991. Therefore, reject H_0. At least one of the distributions differs significantly from at least one of the others.

TRUE-FALSE QUESTIONS

T F   1. Nonparametric tests are generally more powerful than parametric tests.

T F   2. Anytime it is appropriate to use a nonparametric statistic it is appropriate to use a parametric statistic.

T F   3. A χ^2 test can only be applied to nominally scaled variables.

T F   4. In general, nonparametric tests have fewer requirements or assumptions about population characteristics than parametric tests do.

T F   5. As a general rule an investigator should use parametric tests whenever possible to help minimize the probability of making a Type II error.

T F   6. In a single variable χ^2 experiment there are N - 1 degrees of freedom.

T F   7. χ^2 is basically a measure of the overall discrepancy between f_e and f_o.

T F   8. In any specific case f_o - f_e should equal zero if H_0 is true.

T F   9. To use the χ^2 test, the categories in the contingency table must always be mutually exclusive.

T F   10. The value of f_o can be found by multiplying the marginals for that cell and dividing by N.

T F   11. If χ^2 is negative then H_0 must be false.

T F   12. In a 2 x 2 table there are (r - 1)(c - 1) degrees of freedom.

T F   13. The Kruskal-Wallis test is a nonparametric alternate for parametric one-way ANOVA, independent groups design.

T F   14. In general, the Kruskal-Wallis test is as powerful as parametric ANOVA.

T F   15. The theoretical sampling distribution of χ^2 assumes that the distribution is discrete.

T F   16. In order to properly use the χ^2 test each cell should have a value of fe equal to or greater than 10.

T F   17. The sampling distribution of χ^2 is normally distributed.

T F   18. The Wilcoxin signed ranks test is used only with ordinal data.

T F   19. The sign test is less powerful than the Wilcoxin signed ranks test.

T F   20. Both the χ^2 test and the Wilcoxin signed ranks test are one-tailed tests.

T F   21. If the ranking has been done correctly for the Wilcoxin signed ranks test, the sum of the unsigned ranks should equal n(n+1)/2.

T F   22. If $T_{obt} \geq T_{crit}$, reject H_0.

T F   23. When using the Wilcoxin signed ranks test, if the raw scores are tied, the scores are disregarded and N is reduced by 1.

T F   24. The proper use of the Wilcoxin signed ranks test requires that both the raw scores and the difference between the raw scores be of at least ordinal scaling.

T F   25 The Kruskal-Wallis test is used with a correlated groups design.

T F   26 The Kruskal-Wallis test analyzes the difference between sample means.

T F   27 Both the Mann-Whitney U test and the Kruskal-Wallis test analyze differences between sums of ranks.

Answers: 1. F **2.** F **3.** F **4.** T **5.** T **6.** F **7.** T **8.** F **9.** T **10.** F **11.** F **12.** T **13.** T **14.** F **15.** F **16.** T **17.** F **18.** F **19.** T **20.** F **21.** T **22.** F **23.** T **24.** T **25.** F **26.** F **27.** T

SELF-QUIZ

1. The χ^2 test can be used for variables with _____ scaling as long as the categories are mutually exclusive.

   a. nominal
   b. ordinal
   c. interval
   d. ratio
   e. all the above

2. The _____ test is the most powerful test for a repeated measures design.

   a. sign
   b. t
   c. Wilcoxin signed ranks
   d. all the tests are equally powerful

3. Which of the following tests are parametric statistical tests:

   a. sign test
   b. chi-square test
   c. Wilcoxin signed ranks test
   d. none of the above

4. If an experiment using frequency data tested the preference for 6 brands of soup, there would be _____ degrees of freedom.

   a. 1
   b. N - 1
   c. 5
   d. 6

5. The value of χ^2_{obt} for the table below is _____. (Assume equal probabilities for f_e in each cell.)

12	20	16	18	30

a. 96.00
b.   9.42
c.   8.13
d. 13.28

6. The value of χ^2_{crit} for the data in question 5 with $\alpha = .01$ is _____.

a.   6.635
b. 15.086
c. 13.277
d. 11.668

7. The value of f_e for the cell in row W, column A is _____.

	A	B	C
W	16	25	64
X	28	32	30
Y	19	10	42

a. 24.9
b.   3.2
c. 16.0
d. 63.0

8. The value of χ^2_{obt} for the table in problem 7 is _____.

a. 19.01
b. 21.38
c. 24.87
d. 16.82

9. The value of χ^2_{crit} for the table in problem 7 with $\alpha = .05$ is _____.

    a. 13.277
    b.   7.779
    c.   3.841
    d.   9.488

10. For the following table there are _____ degrees of freedom.

	X	Y
A	60	30
B	18	40

    a. k - 1
    b. 1
    c. 2
    d. 4

11. The value of χ^2_{obt} for the table in problem 10 is _____.

    a. 47.43
    b. 17.96
    c.   4.07
    d.   9.71

12. For a low value of df the χ^2 distribution is _____.

    a. normally distributed
    b. positively skewed
    c. negatively skewed
    d. none of the above

13. The value of T_{obt} for the following data is _____.

Subject	Pre	Post
1	100	98
2	62	80
3	80	75
4	75	74
5	90	80
6	85	71

a. 21
b. -6
c. 15
d. 6

14. If there are 16 subjects in a repeated measures design then the sum of the unsigned ranks equals _____.

a. 136
b. 68
c. 272
d. 32

15. If T_{obt} = 12 and T_{crit} = 10, one would _____.

a. reject H_0
b. retain H_0
c. accept H_0
d. reject H_1

Questions 16 - 21 pertain to the following information.

In Chapter 16, we presented the data from an independent groups design and asked if it was appropriate to use parametric ANOVA. The data are presented again below. The correct answer was that it was not appropriate to use parametric ANOVA because of unequal n's and homogeniety of variance assumption violation.

Group 1	Group 2	Group 3
6	5	20
7	4	92
9	4	68
8	3	31
2	1	4
3		10
		82
		212

16 Is it possible to analyze the data with an alternate test?

    a. yes
    b. no

17 If your answer to 16 is yes, what is the name of the test?

    a. t test for independent groups
    b. F test
    c. Kruskal-Wallis
    d. Mann-Whitney U test

18 $H_{obt} = $ _____ .

    a. 10.25
    b. 10.63
    c. 15.96
    d.   5.96

19 What are the df?

    a. 1
    b. 2
    c. 3
    d. need more information

20 Using $\alpha = .05$, $H_{crit} = $ _____ .

    a. 3.841
    b. 7.815
    c. 5.991
    d. 7.824

21 What do you conclude?

    a. retain H_0. There is no difference in the populations.
    b. accept H_0. There is no difference in the populations.
    c. reject H_0. At least one of the population means differs from at least one of the others.
    d. reject H_0. At least one of the distributions differs from at least one of the others.

Answers: 1. e   **2.** b   **3.** d   **4.** c   **5.** b   **6.** c   **7.** a   **8.** b   **9.** d   **10.** b
11. b   **12.** b   **13.** d   **14.** a   **15.** b   **16.** a   **17.** c   **18.** b   **19.** b   **20.** c
21. d.

20 REVIEW OF INFERENTIAL STATISTICS

This chapter presents a concise review of hypothesis testing and inferential statistics. To outline or review the concepts covered in this chapter would be redundant. The exercises and questions to follow can serve as a self-review over this material. You might wish to use it as a practice examination. Be sure to follow good procedure by using the most powerful test that is appropriate for the information presented.

EXERCISES

1. Name two different pieces of information given by the sampling distribution of a statistic.

2. What are the three steps in empirically generating a sampling distribution?

3. In an ideal experiment, what should be the magnitudes of power and alpha?

4. Why is the z test for independent groups rarely used?

5. A neurologist believes that certain chemical toxins can affect the ability of nerves to conduct impulses. Eleven experimental animals have their nerve conduction velocity measured before and after exposure to the toxin. The following results are observed. The measures are in meters per second.

Animal	Before Exposure	After Exposure
1	45	40
2	42	36
3	39	40
4	50	48
5	46	42
6	44	38
7	38	38
8	41	49
9	40	42
10	47	42
11	46	41

   a. State the nondirectional alternative hypothesis.
   b. State H_0.
   c. What do you conclude using $\alpha = .01_{2\,tail}$?
   d. What type error might you be making?

6. Analyze the data from problem 5 using the most powerful nonparametric test appropriate for this design. What do you conclude?

7. A study was performed to see if cars actually get the gas mileage the EPA says. A sample of six cars was randomly selected from a manufacturer, tuned to specifications and then mileage tested using identical procedures as the EPA uses. The mean miles per gallon (mpg) from the sample was 27.8 mpg with a standard deviation of 1.9 mpg. The EPA states that the mileage should be 30.0 mpg. What would you conclude using $\alpha = .05_{2\,tail}$? Assume a normally distributed population.

8. An insurance company wanted to know if a high school driver's education class reduced the number of accidents 16 year old drivers had. Eighteen high schools were selected. The number of accidents per 100 students was measured among students who had and had not taken driver's education. The results are shown below. Assume that the groups are independent.

With Drivers Education	Without Drivers Education
8	9
6	11
7	8
2	9
3	8
4	16
0	10
9	7
7	9

  a. State the directional alternative hypothesis.
  b. State the null hypothesis.
  c. What do you conclude using $\alpha = .01_{1\ tail}$?

9. Analyze the data from problem 8 using the most powerful nonparametric test appropriate. What do you conclude?

10. In an effort to determine if past marital history is related to present marital adjustment, a retrospective survey of public records reveals the following information for a 10 year period. Out of 400 males selected for the sample, 150 had been married 2 or more times at the outset of the 10 year period. Of these 150, 100 were divorced again during the 10 years. Of the 250 males who were in their first marriage at the outset of the 10 year period, 123 got divorced during the following 10 years. The data are summarized in the table below.

Divorced in 10 year period

	No	Yes
One Marriage	127	123
Two or More Marriages	50	100

Are past marital history and future marital success related? Use $\alpha = .01$. Round f_e to one decimal place.

11. A study was conducted to determine if a correlation exists between the unemployment rate in a community and the number of suicides over a period of 10 years. The correlation coefficient for the 10 pairs of scores was calculated with a resulting $r_{obt} = .656$.

a. State the null hypothesis.

b. Is this value of r_{obt} significant using $\alpha = .01_{2 \text{ tail}}$?

12. A consumer testing laboratory wants to know if a sample of 20 boxes of breakfast cereal with a mean weight of 15.6 ounces could reasonably have been drawn from a population which is supposed to contain 16 ounces with a standard deviation of 0.6 ounces. What should the investigator conclude using $\alpha = .05_{2 \text{ tail}}$? Assume the population is normally distributed.

13. A manufacturer of feed grain has three different products that it wishes to test. The question of interest is whether groups of test animals live longer eating certain products. The following data are the life span in months of groups of test chickens.

Product A	Product B	Product C
40	36	42
39	35	40
34	42	43
38	38	39
42	46	47
37	37	38
41	40	44
43	46	47

a. What is the alternative hypothesis?
b. What is the null hypothesis?
c. Using $\alpha = .05$, what do you conclude?

14. A high school senior wants to know if studying for the Scholastic Aptitude Test (SAT) produces higher grades on the test. She surveyed two groups of students who had taken the test the year before. She took data from 20 students all of whom had about the same high school GPA. The following data were gathered on the two groups of students.

No Study	Study
620	750
580	685
540	770
720	600
710	500
400	790
456	490
500	540
525	645
600	500

a. State the nondirectional null hypothesis.
b. What do you conclude using the t test and $\alpha = .05_{2\,tail}$?
c. If you felt the t test was inappropriate because you believed the SAT scores were only ordinal in scaling, what other test would be appropriate?
d. What would you conclude using that test and $\alpha = .05_{2\,tail}$?

15. A neuropsychologist knows that in the unanaesthetized population the mean reaction time is 285 milliseconds. The population is normally distributed. A sample of 17 subjects who have undergone general anaesthesia but no surgical procedure are given this same task 24 hours after receiving the anaesthesia. The reaction time for this random sample of subjects was $\overline{X} = 327$ msec and s = 22 msec. Does the anaesthesia affect performance on this task? Use $\alpha = .05_{2\,tail}$.

16. A cosmetic manufacturer wants to add a new perfume to its line of products. Five test fragrances are presented to a sample of women. The order is randomized. The preferences are shown in the table below.

Test Fragrance

A	B	C	D	E	
16	21	38	30	9	114

a. What is the nondirectional alternative hypothesis?
b. What do you conclude using $\alpha = .05$?

17. A nutritionist believes that large doses of vitamin C will affect the number and severity of the colds a person gets during a winter. Two groups of volunteers are given either a placebo or 1000 mg of vitamin C per day. At the end of the winter each subject was asked to rate their health with regard to colds on a 20 point ordinal rating scale. The

results are shown below. Higher scores represent better health. Assume the scores come from a highly skewed population.

Placebo	Vitamin C
16	12
5	6
11	18
14	14
19	4
10	15
16	17
3	20

   a. State the nondirectional alternative hypothesis.
   b. State H_0.
   c. What do you conclude using $\alpha = .05_{2 \text{ tail}}$?
   d. What type of error might you be making?

18. A health educator believes that there is evidence to support the hypothesis that there is a positive relationship between the amount someone is over their ideal body weight and their serum cholesterol. She obtains the following data.

Excess Weight (lbs)	Serum Cholesterol (mg/100 ml)
1	125
5	100
8	130
10	200
15	280
30	320
60	380
20	250
12	260

   a. What is the value of r_{obt}?
   b. Is r_{obt} significantly more positive than 0 using $\alpha = .05_{1 \text{ tail}}$?
   c. What type error might we be making?

19. A behavioral scientist wishes to compare the effectiveness of five different weight loss programs. Six subjects are randomly assigned to each of the programs and the weight loss per month is recorded.

Program

1	2	3	4	5
6	10	7	14	0
8	11	6	16	5
7	12	8	8	6
5	11	10	15	2
3	8	8	15	2
9	5	13	18	3

a. State H_0.
b. What is the value of F_{obt}?
c. What do you conclude using $\alpha = .01$?
d. What does a planned comparison between groups 1 and 4 reveal? (Use $\alpha = .01_{2 \text{ tail}}$)
e. Compare groups 2 and 5 using the most conservative post hoc test you know using $\alpha = .01$.

20. A pilot study was conducted in a school to see if a cancer information curriculum was effective in conveying new information to students. A pre and post test were administered to a small sample of students. The data are shown below.

Student	A	B	C	D	E	F	G	H	I	J
Pre Test	56	58	73	65	60	62	64	68	70	73
Post Test	75	81	72	80	69	68	69	64	69	89

a. What do you conclude when analyzing these data using the sign test with $\alpha = .05_{1 \text{ tail}}$?
b. What do you conclude using $\alpha = .05_{1 \text{ tail}}$ and the Wilcoxin signed ranks test?
c. What do you conclude using $\alpha = .05_{1 \text{ tail}}$ and the t test?
d. How do you explain these differing results?
e. What is the probability of detecting an effect represented by $P_{real} = .70$ using the sign test and $\alpha = .05_{1 \text{ tail}}$?

21. What is the value of t_{obt} for evaluating the significance of $r = .83$ for $N = 15$ pairs of scores?

22. A vetinarian believes his diet formula fed to dogs will produce healthier, less fat animals. His records collected on thousands of dogs show a mean of 22% fat with a standard deviation of 5.4. The data is normally distributed.

    a. If he tests his formula on a sample of 30 dogs, what is the power of this experiment to detect an effect of the diet such as to produce a mean decrease in fat of 2%? Use $\alpha = .05_{1\ tail}$.

    b. What number of dogs should he use to achieve a power of .8500 to detect the effect postulated in 22a?

23. A study was conducted to determine if people's performance on a standard task is related to their level of anxiety. A group of 1,018 subjects was randomly assigned to one of three conditions designed to create a low, moderate or high level of anxiety. Their performance on the standard task was rated as poor, fair, or good by an independent observer. The following results were obtained.

		Poor	Fair	Good
	Low	130	100	80
Anxiety Level	Moderate	47	193	162
	High	170	80	56

Performance Rating

    a. State H_1.
    b. State H_0.
    c. What do you conclude using $\alpha = .01$? Round f_e to one decimal place accuracy.

24. The value of s_W^2 from a one-way ANOVA experiment which resulted in a significant overall F was 21.76. There were equal n's in each of four groups. The total degrees of freedom equaled 27. The means of the groups are as follows:

$$\bar{X}_1 = 17.2 \quad \bar{X}_2 = 14.8 \quad \bar{X}_3 = 21.3 \quad \bar{X}_4 = 24.0$$

    a. Using the Newman-Keuls test and $\alpha = .05$, is group 1 different from group 3?
    b. Using the Newman-Keuls test and $\alpha = .05$, is group 2 different from group 3?

c. Using the HSD test, is group 2 different from group 3?

d. Explain the difference between the conclusions of b and c.

25. Is it reasonable to conclude that the following random sample of IQ scores could have been drawn from a normally distributed population of scores with $\mu = 100$? Use $\alpha = .05_{2\ tail}$.

IQ Scores
107
106
92
103
110
105
100
116

26. A sleep researcher believes that when a person is sleep deprived for over 20 hours he becomes depressed. A group of 12 subjects is given a depression rating scale with ordinal scaling properties before and after 20 hours of sleep deprivation. The data are:

Subject	Before	After
1	56	65
2	70	71
3	40	43
4	37	59
5	65	70
6	50	70
7	55	61
8	72	65
9	68	64
10	66	80
11	61	69
12	57	71

Higher scores indicate increased depression.

a. State the nondirectional alternative hypothesis.

b. State the null hypothesis.

c. What do you conclude using $\alpha = .05$ Assume population depression scores are J-shaped

27. A study evaluates the effectiveness of three types of therapy in producing self-reliance. Fifteen voluntaires are randomly assigned, five to each type of therapy. A questionaire measuring self-reliance is administered after therapy is completed. The following results are obtained. Higher scores indicate greater self-reliance.

	Therapy	
Type A	Type B	Type C
10	9	16
11	13	18
12	14	19
15	8	17
7	14	15

a. Assume the data meet the assumptions of parametric ANOVA. What do you conclude, using $\alpha = .05$?

b. Assume the data seriously violate the assumptions of parametric ANOVA. Use another test to analyze the data. Again use $\alpha = .05$.

28. The researcher who did the study in question 27 has a hunch that the three types of therapy may effect males and females differently. Therefore a second study is undertaken to replicate the first study and in addition to see if there is a male/female difference in the effects of the therapy types. An independent groups design is conducted with 5 subjects per cell. The same questionaire is used. The following data are collected.

		Therapy	
Gender	Type A	Type B	Type C
Female	10	9	16
	11	13	18
	12	14	19
	15	8	17
	7	14	15
Male	9	8	18
	12	12	18
	12	15	17
	14	9	15
	8	14	20

What do you conclude, using $\alpha = .05$?

Answers: 1. First, all the values the statistic can take and second, the probability of getting each of those values if chance alone is responsible or sampling is random from the Null Hypothesis Population.

2. Step 1: all possible different samples of size N that can be formed from the population are determined. Step 2: the statistic for each of the samples is calculated. Step 3: the probability of getting each value of the statistic is calculated under the assumption that sampling is random from the Null Hypothesis Population.

3. Ideally, power would be high (power = 1.0000) and alpha would be low ($\alpha = 0$).

4. Because the experimenter rarely knows the population parameters μ and σ.

5a. H_1: Exposure to the toxin affects nerve conduction velocity.   **5b.** H_0: Exposure to the toxin has no effect on nerve conduction velocity.   **5c.** $t_{obt} = 1.51$; $t_{crit} = \pm 3.169$. Therefore, retain H_0.   **5d.** Type II.

6. Using the Wilcoxin signed ranks test, $T_{obt} = 13.5$; $T_{crit} = 3$; therefore retain H_0.

7. $t_{obt} = -2.84$; $t_{crit} = \pm 2.571$. Therefore reject H_0. It is not reasonable to assume that this sample was drawn from a population with $\mu = 30$.

8a. H_1: driver's education classes reduce the number of accidents in which students are involved.   **8b.** H_0: driver's education classes do not reduce the number of accidents in which students are involved.   **8c.** $t_{obt} = -3.40$; $t_{crit} = -2.583$. Therefore, reject H_0.

9. Using the Mann-Whitney U test, $U_{obt} = 7.5$ and $U'_{obt} = 73.5$; $U_{crit} = 14$. Therefore, reject H_0.

10. $\chi^2_{obt} = 11.59$; $\chi^2_{crit} = 6.635$. Therefore, reject H_0. Past marital history and current adjustment are not independent.

11a. H_0: the sample is a random sample from a population with $\rho = 0$ and that any correlation in the sample is due to chance alone.   **11b.** $r_{obt} = 0.656$; $r_{crit} = \pm 0.7646$. Therefore, retain H_0.

12. $z_{obt} = -2.98$; $z_{crit} = \pm 1.96$. Therefore, it is not likely that the sample boxes were drawn from a population with the stated parameters (i.e., reject H_0).

13a. H_1: at least one of the products differentially affects life span.   **13b.** H_0: none of the products differentially affects life span.   **13c.** $F_{obt} = 1.79$; $F_{crit} = 3.47$. Therefore, retain H_0.

14a. H_0: studying has no effect on SAT scores.   **14b.** $t_{obt} = -1.25$; $t_{crit} = \pm 2.101$. Therefore, retain H_0.   **14c.** Mann-Whitney U test   **14d.** $U_{obt} = 36$; $U'_{obt} = 64$; $U_{crit} = 23$. Therefore, retain H_0. Notice again that it is unlikely that a nonparametric test will allow us to reject H_0 if a parametric on the same data has not allowed rejection of H_0.

15. $t_{obt} = 7.87$; $t_{crit} = \pm 2.120$. Therefore, reject H_0.

16a. H_0: there is no difference in preference among the test fragrances.   **16b.** $\chi^2_{obt} = 22.93$; $\chi^2_{crit} = 9.488$. Therefore, reject H_0.

17a. H_1: vitamin C affects the health ratings of individuals who take it.   **17b.** H_0: vitamin C has no effect on the health ratings of individuals who take it .   **17c.** We have used the Mann-Whitney U test because the normal assumption for the t test is violated and N is relatively small. In addition, some researchers would also argue that you can't use the t test because it requires interval or ratio data. $U_{obt} = 25.5$; $U_{crit} = 13$. Therefore, retain H_0.   **17d.** Type II.

18a. $r_{obt} = 0.86$ **18b.** $r_{crit} = 0.5822$. Therefore, reject H_0. It seems likely that this sample was not drawn from a population of scores with $\rho = 0$.   **18c.** Type I .

19a. H_0: there is no difference in weight loss between these groups   **19b.** $F_{obt} = 15.47$ **19c.** $F_{crit} = 4.18$. Therefore, reject H_0.   **19d.** $t_{obt} = 5.32$; $t_{crit} = \pm 2.787$. Reject H_0. These groups are not very likely to have been drawn from the same population.   **19e.** Use HSD test. $Q_{obt} = 6.11$; $Q_{crit} = 5.15$. Therefore, reject H_0.

20a. $p = .1719$ and $\alpha = .05$. Therefore, retain H_0.   **20b.** $T_{obt} = 6$, $T_{crit} = 10$. Therefore, reject H_0.   **20c.** $t_{obt} = 2.97$; $t_{crit} = 1.833$. Therefore, reject H_0.   **20d** The sign test was not powerful enough to allow us to reject H_0 **20e.** Power $= 0.1493$.

21. $t_{obt} = 5.37$.

22a. Power $= .6480$ **22b.** $N = 53$.

23a. H_1: anxiety level and performance are not independent.   **23b.** H_0: anxiety level and performance are independent.   **23c.** $\chi^2_{obt} = 161.63$; $\chi^2_{crit} = 13.277$. Therefore, reject H_0.

24a. $Q_{obt} = 2.33$; $Q_{crit} = 2.92$. Therefore, we cannot say groups 1 and 3 are drawn from different populations.   **24b.** $Q_{obt} = 3.69$; $Q_{crit} = 3.53$. Therefore, using this test, we reject H_0.   **24c.** $Q_{obt} = 3.69$; $Q_{crit} = 3.90$. Therefore, retain H_0.   **24d.** HSD is more conservative regarding Type I error probability than Newman-Keuls.

25. $t_{obt} = 1.95$; $t_{crit} = \pm 2.365$. Therefore, retain H_0. It is reasonable that this sample could have been drawn from a population with $\mu = 100$.

26a. H_1: sleep deprivation affects depression rating scores.   **26b.** H_0: sleep deprivation has no effect on depression rating scores.   **26c.** We have used the Wilcoxin signed ranks test instead of the t test because the population scores are not normally distributed and N is small. In addition, many researchers would rule out use of the t test because the data are only of ordinal scaling. $T_{obt} = 9$; $T_{crit} = 13$. Therefore, reject H_0.

27a. $F_{obt} = 8.49$; $F_{crit} = 3.74$. Therefore reject H_0.   **27b.** $H_{obt} = 9.04$; $H_{crit} = 5.991$. Therefore reject H_0.

28. F_{obt} (row) $= 0.05$; $F_{crit} = 4.26$. Retain H_0. F_{obt} (column) $= 19.14$; $F_{crit} = 3.40$. Reject H_0. F_{obt} (row $\times$ column) $= .05$; $F_{crit} = 3.40$. Retain H_0. There is a significant main effect for the therapy types, and no other significant effects. This study replicates the previous finding and extends it to males.

TRUE-FALSE QUESTIONS

T F   1. H_0 and H_1 must be mutually exclusive and exhaustive.

T F   2. If H_0 is true then H_1 must also be true.

T F   3. H_0 always states that the dependent variable has no effect on the independent variable.

T F   4. In a replicated measures design the Null Hypothesis Population has a population of difference scores with $\mu_D = 0$.

T F   5. The sampling distribution of a statistic can be generated by either an empirical or theoretical approach.

T F   6. The critical region for rejection is all the area under the curve.

T F   7. The critical value of a statistic depends on the alpha level.

T F   8. If one retains H_0 and H_0 is true, one has made a Type I error.

T F   9. $\alpha + \beta = 1$

T F   10. As power increases, alpha decreases.

T F   11. If in reality H_0 is true, then increasing power increases the probability of making a Type I error.

T F   12. As α increases, β decreases.

T F 13. With single sample experimental designs, one or more of the population parameters must be known.

T F 14. The t and z tests evaluate the effect of the independent variable on the sample mean while the ANOVA evaluates the effect of the independent variable on the sample variance.

T F 15. The sampling distribution of the mean has a mean of $\mu_{\overline{X}} = \mu$.

T F 16. The mean of all sampling distributions equals 0.

T F 17. If the raw score population is normally distributed, then so is the corresponding sampling distribution of the mean regardless of sample size.

T F 18. If the raw score population is not normally distributed, then the corresponding sampling distribution of the mean will not be normally distributed regardless of sample size.

T F 19. For a t test with single sample there are N - 1 degrees of freedom.

T F 20. The Wilcoxin signed rank test takes into account the magnitude and direction of the raw scores.

T F 21. To utilize the sign test, P must equal Q.

T F 22. Ties are not counted in the sign test.

T F 23. The Mann-Whitney U test requires equal n's in both groups.

T F 24. $F_{obt} = SS_B / SS_W$ (one-way ANOVA).

T F 25. If there are k treatments and k means between $\overline{X}_i$, and $\overline{X}_j$, then Q_{crit} will be the same for the HSD test and Newman-Keuls test when comparing $\overline{X}_i$ and $\overline{X}_j$.

T F 26. In the analysis of variance, it is not necessary for the sample variances to be homogeneous since this is what the statistic analyzes.

T F 27. The χ^2 test is only for use with nominal data.

T F 28. There are (r - 1) + (c - 1) degrees of freedom in the χ^2 test.

T F 29. The χ^2 test can be used in a repeated measures design.

T F 30. For the proper use of the χ^2 test, f_e should equal f_o.

T F 31. To properly use the χ^2 test, the data must be reduced to mutually exclusive categories and appropriate frequencies.

T F 32. Power and sample size are directly related, while power and sample variance are inversely related.

T F 33. It is not possible to test the null hypothesis that P = .50.

T F 34 The Kruskal-Wallis test is a nonparametric test used in conjunction with a one-way independent groups design.

T F 35 In two-way ANOVA, we test for the main effects of two variables and their interaction.

T F 36 A significant interaction means the effects of one of the variables are not the same at all levels of the other variable.

Answers: 1. T   **2.** F   **3.** F   **4.** T   **5.** T   **6.** F   **7.** T   **8.** F   **9.** F   **10.** F
11. F   **12.** T   **13.** T   **14.** F   **15.** T   **16.** F   **17.** T   **18.** F   **19.** T   **20.** T
21. F   **22.** T   **23.** F   **24.** F   **25.** T   **26.** F   **27.** F   **28.** F   **29.** F   **30.** F
31. T   **32.** T   **33.** F   **34.** T   **35.** T   **36.** T.

SELF-QUIZ

1. The value of z_{crit} for $\alpha = .02_{2\ tail}$ is _____.

   a. ±1.96
   b. ±1.64
   c. ±2.33
   d. ±2.58

2. The value of t_{crit} for $\alpha = .05_{2\ tail}$ in a repeated measures experiment with N = 20 subjects is _____.

   a. ±2.093
   b. ±1.960
   c. ±2.861
   d. ±2.086

3. The value of r_{crit} for $\alpha = .01_2$ tail with N = 15 pairs of observations is _____.

    a. ±0.5923
    b. ±0.6055
    c. ±0.6226
    d. ±0.6411

4. The F distribution is _____.

    a. positively skewed
    b. has no negative values
    c. has a median approximately equal to 1
    d. all the above

5. There are _____ degrees of freedom for s_w^2

    a. k - 1
    b. N - k
    c. N - 1
    d. N x k

6. The area under the critical region for rejection when $\alpha = .05_2$ tail is _____.

    a.   .05
    b.   .025
    c.   .95
    d. 1.00

7. Which of the following levels of alpha is illegal?

    a. .01
    b. .05
    c. .10
    d. none of the above, all are permissible depending on the situation

8. If power equals 0.860, then β equals _____.

    a. 1.860
    b. 0.860
    c. 0.140
    d. 0.950

9. If $\mu = 37$, $\sigma = 56$, and $X = 42.6$, then the values of z_{obt} using the z test for single samples equals _____.

    a. 1.00
    b. -1.00
    c. 0.68
    d. cannot be determined from information given

10. Consider an independent groups design with $k = 2$ and $s_1^2 = 0.75$ and $s_2^2 = 12.6$ and $n_1 = n_2 = 8$. What test should you use to analyze these data?

    a. t test
    b. F test
    c. Mann-Whitney U test
    d. any of the above

11. Which of the following do (does) not require equal n's in each group?

    a. t test for independent groups
    b. ANOVA
    c. Mann-Whitney U test
    d. none of the above require equal n's
    e. a and c

12. If $t_{obt} = 2.86$ for an independent groups design, then F_{obt} for the same data equals _____.

    a. 8.18
    b. 1.69
    c. 2.86
    d. cannot be determined from information given

13. Consider the following data from an independent groups design.

Condition 1	Condition 2
16	17
31	25
19	24
20	31
22	30
19	29
25	24

What is the value of t_{obt}?

   a. -2.34
   b. -1.53
   c.   2.34
   d. -1.69

14. If the data were ordinal what would the value of U_{obt} be from the data in problem 13?

   a. 35
   b. 17
   c. 14
   d. 49

15. The following data were gathered to assess the impact of a speed reading course. A repeated measures design was employed. The dependent variable is words per minute.

Before	After
300	800
240	650
350	900
400	500
500	300
200	201
375	390
550	700
600	630

What is the value of t_{obt}?

   a. -4.08
   b.   1.46
   c. -2.02
   d. -3.52

16. If the data in problem 15 were analyzed using the Wilcoxin signed rank test, the value of T_{obt} would be _____.

   a.   7
   b. 29
   c. 36
   d.   5

17. If one were using the sign test to analyze the results of problem 15, the probability of getting the results observed or results more extreme would be _____.

a. 0.1406
b. 0.0392
c. 0.0703
d. 0.0899

18. What is the power of the sign test using $\alpha = .05_{2\ tail}$ on the data in problem 15 if $P_{real} = .80$?

a. 0.4362
b. 0.3679
c. 0.0196
d. 0.0392

19. What is the value of F_{obt} for the following data from an experiment with $k = 3$ groups and ratio scaled variables.

Group A	Group B	Group C
1.8	1.9	0.4
1.6	2.0	0.2
1.0	2.0	0.8
1.3	1.6	0.3

a.   2.20
b. 29.05
c.   6.49
d.   5.39

20. Referring to the data in problem 19, what is the value of Q_{crit} for comparing Group A and Group C using $\alpha = .05$ and the Newman-Keuls test?

a. 4.60
b. 3.95
c. 5.43
d. 3.20

21. Referring to the data in problem 19, what is the value of Q_{crit} for comparing Group A and Group C using $\alpha = .05$ and the HSD test?

    a. 4.60
    b. 3.95
    c. 5.43
    d. 3.20

22. What is the value of Q_{obt} for a post hoc comparison of groups A and C from problem 19?

    a.   7.26
    b.   0.14
    c.   1.00
    d. 10.88

23. A scientist asks a group of 75 other scientists which of three professional organizations they believe has done the most to advance the cause of free scientific investigation. The following data were obtained.

Organization

A	B	C	
30	25	20	75

What is the value of χ^2_{obt}

    a. 0.60
    b. 1.00
    c. 2.00
    d. 9.00

24. The following data are obtained on whether there is a relationship between type of physician and treatment preference for cancer. Round f_e to one decimal place accuracy.

Treatment Preference

	Surgery	Chemo-therapy	Both
Surgeon	62	27	46
Internist	37	89	47

What is the value of χ^2_{obt}?

a. 3.79
b. 16.03
c. 21.06
d. 35.21

25. What is the value of χ^2_{obt} for the following table which relates sex to birth control preference? Round f_e to one decimal place accuracy.

	Oral Contra-ceptive	Vasec-tomy
Male	50	10
Female	20	21

a. 12.04
b. 13.67
c. 8.71
d. 1.27

26. What is the value of t_{obt} for the test of significance of r if $r_{obt} = .58$ and $N = 30$

a. 3.58
b. 6.30
c. 3.77
d. 2.94

27. The following data are collected in a two-way independent groups design. Assume the data are ratio scaled and come from normal populations.

Variable A	Variable B Level 1	Level 2
Level 1	2	7
	1	5
	5	8
	3	6
Level 2	9	1
	7	1
	8	2
	6	4

Which of the following conclusions is correct? Use $\alpha = .05$.

a. There are no significant effects
b. There is a significant main effect for variable A, and no other significant effects
c. There is a significant main effect for variable B, and no other significant effects
d. There is a significant interaction effect and no other significant effects

Answers: 1. c   **2.** a   **3.** d   **4.** d   **5.** b   **6.** a   **7.** d   **8.** c   **9.** d   **10.** c
11. d   **12.** a   **13.** b   **14.** c   **15.** c   **16.** d   **17.** b   **18.** a   **19.** b   **20.** d
21. b   **22.** a   **23.** c   **24.** d   **25.** b   **26.** c.   **27.** d.